# COLLEGES OF AGRICULTURE AT THE LAND GRANT UNIVERSITIES

**A PROFILE**



Committee on the Future of the Colleges of Agriculture in the Land Grant University System
Board on Agriculture
National Research Council

NATIONAL ACADEMY PRESS
Washington, D.C. 1995

**NATIONAL ACADEMY PRESS 2101 Constitution Avenue, NW Washington, DC 20418**

NOTICE: The project that is the subject of this report was approved by the Governing Board of the National Research Council, whose members are drawn from the councils of the National Academy of Sciences, the National Academy of Engineering, and the Institute of Medicine. The members of the committee responsible for the report were chosen for their special competencies and with regard for appropriate balance.

This report has been reviewed by a group other than the authors according to procedures approved by a Report Review Committee consisting of members of the National Academy of Sciences, the National Academy of Engineering, and the Institute of Medicine.

This study was supported by the Cooperative State Research, Education, and Extension Service of the U.S. Department of Agriculture, under Agreement No. 93-COOP-2-8575 and by the W. K. Kellogg Endowment Fund of the National Academy of Sciences and the Institute of Medicine. The Economic Research Service of the U.S. Department of Agriculture also provided support. Any opinions, findings, conclusions, or recommendations expressed in this publication are those of the authoring committee and do not necessarily reflect the views of the sponsors.

The National Academy of Sciences is a private, nonprofit, self-perpetuating society of distinguished scholars engaged in scientific and engineering research, dedicated to the furtherance of science and technology and to their use for the general welfare. Upon the authority of the charter granted to it by the Congress in 1863, the Academy has a mandate that requires it to advise the federal government on scientific and technical matters. Dr. Bruce M. Alberts is president of the National Academy of Sciences.

The National Academy of Engineering was established in 1964, under the charter of the National Academy of Sciences, as a parallel organization of outstanding engineers. It is autonomous in its administration and in the selection of its members, sharing with the National Academy of Sciences the responsibility for advising the federal government. The National Academy of Engineering also sponsors engineering programs aimed at meeting national needs, encourages education and research, and recognizes the superior achievements of engineers. Dr. Harold Liebowitz is president of the National Academy of Engineering.

The Institute of Medicine was established in 1970 by the National Academy of Sciences to secure the services of eminent members of appropriate professions in the examination of policy matters pertaining to the health of the public. The Institute acts under the responsibility given to the National Academy of Sciences by its congressional charter to be an adviser to the federal government and, upon its own initiative, to identify issues of medical care, research, and education. Dr. Kenneth I. Shine is president of the Institute of Medicine.

The National Research Council was organized by the National Academy of Sciences in 1916 to associate the broad community of science and technology with the Academy's purposes of furthering knowledge and advising the federal government. Functioning in accordance with general policies determined by the Academy, the Council has become the principal operating agency of both the National Academy of Sciences and the National Academy of Engineering in providing services to the government, the public, and the scientific and engineering communities. The Council is administered jointly by both Academies and the Institute of Medicine. Dr. Bruce M. Alberts and Dr. Harold Liebowitz are chairman and vice chairman, respectively, of the National Research Council.

**Library of Congress Cataloging-in-Publication Data**

Colleges of agriculture at the land grant universities : a profile / Committee on the Future of Land Grant Colleges of Agriculture, Board on Agriculture, National Research Council.

p. cm.

Includes bibliographical references (p. ) and index.

ISBN 0-309-05295-5 (alk. paper)

1. Agricultural colleges—United States. 2. Agricultural education—United States. 3. State universities and colleges— United States. I. National Research Council (U.S.). Committee on the Future of Land Grant Colleges of Agriculture.

S533.C695 1995

630'.71'173—dc20 95-31198

CIP



Printed in the United States of America

## COMMITTEE ON THE FUTURE OF THE COLLEGES OF AGRICULTURE IN THE LANDGRANT UNIVERSITY SYSTEM

ANTHONY S. EARL, *Chair*, Quarles and Brady Law Firm, Madison, Wisconsin
R. LEE BALDWIN, University of California, Davis
JOHN C. GORDON, Yale University
GORDON E. GUYER, Michigan Department of Agriculture
FRED HARRISON, JR., Fort Valley State College
EDWARD A. HILER, Texas A&M University
MARLYN L. JORGENSEN, Jorg-Anna Farms, Garrison, Iowa
DARYL B. LUND, Cornell University
THOMAS F. MALONE, The Sigma Xi Center, Research Triangle Park, North Carolina
MORTIMER H. NEUFVILLE, University of Maryland, Eastern Shore
ELIZABETH D. OWENS, ISK Biosciences Corporation, Mentor, Ohio
C. ALAN PETTIBONE, Washington State University
ALLEN ROSENFELD, Public Voice for Food and Health Policy, Washington, D.C.
CHARLES F. SAUL, Agway, Inc., Syracuse, New York (Retired)
G. EDWARD SCHUH, University of Minnesota
GEORGE E. SEIDEL, JR., Colorado State University
JO ANN DOKE SMITH, Smith Associates, Irving, Texas
KATHERINE R. SMITH, Henry A. Wallace Institute for Alternative Agriculture, Greenbelt, Maryland
JAMES B. WYNGAARDEN, Duke University
ELISABETH A. ZINSER, University of Kentucky
JAMES J. ZUICHES, Washington State University

**Staff**

NICOLE BALLENGER, *Project Director*
CARLA CARLSON, *Director of Communications*
DIBY KOUADIO, *Research Associate*
JANET OVERTON, *Editor*
VIOLA HOREK, *Project Assistant*

# BOARD ON AGRICULTURE

# PREFACE

The colleges of agriculture have a proud history. They have contributed to remarkable advances in both farming productivity and agricultural science and technology, which in turn have contributed to the growth of the U.S. economy and the well-being of consumers the world over. It is because of this success that at the beginning of 1995 colleges of agriculture, now more than a century old, still dot the U.S. landscape. They can be found at land grant colleges and universities in almost every state and territory of the nation.

The colleges of agriculture are confronting significant challenges to their future. These challenges are the result of the changing role of farming in the United States and the corresponding changes in the interests of U.S. citizens in agriculture—that is, in the food, fiber, and natural resource complex. In 1862 when the colleges of agriculture were being instituted, farmers comprised more than one-half of the nation's working population. Agriculture-related interests of farmers, consumers, and other groups probably coincided reasonably well. Today farming represents only a small share of the U.S. economy, but the entire agricultural complex (including food and fiber production, processing, and marketing) is of significant economic importance and increasingly driven by consumer wants and concerns. Diverse groups, many of them now urban and suburban, are interested in how the workings of the agricultural complex affect nutrition and health, consumer and worker safety, convenience, the environment, and animal welfare and thus have a stake in research and education at the colleges of agriculture. The colleges' challenges are compounded by developments in science and its infrastructure—developments that are changing the research relationships among universities, government, and private industry—and by competing demands on limited budgets for publicly funded research and education.

The colleges of agriculture in the land grant university system are truly public institutions created and shaped by federal legislation that endowed them with three functions—to instruct students, to perform agriculture-related research, and to provide extension services to farmers; and these functions are still largely supported by state and federal funds. The colleges' contributions—to agricultural output in general and farming productivity in particular through research and extension and to the education of those who farm, manage agribusinesses, lead communities, teach, and conduct scientific research—are the result of the public's investments. It is thus appropriate, and in fact important, to expect accountability with respect to the system's use of public dollars and to evaluate the evolution of the colleges' work in relation to changing public needs and priorities.

This publication is the first of two volumes by the Committee on the Future of the Colleges of Agriculture in the Land Grant University System, convened by the National Research Council's Board on Agriculture. The committee's charge is to assess the adaptation of the land grant colleges to the public's changing needs and priorities and to recommend public policy and institutional change that can enhance the colleges' role in serving the national interest. The committee recognizes that its work must be underpinned by a solid understanding of the colleges' roots; evolution; activities in relation to national, state, and local needs; and potential for change. The committee, in addition to capitalizing on its members' diverse backgrounds and expertise, has been garnering knowledge through a two-stage process. The first stage is an assessment of data and information that describes the colleges, their activities, and their operating context. The second stage is an up-close examination of the interface between college activities and public needs through a series of forums held at land grant colleges in Connecticut, Missouri, New Mexico, South Carolina, and South Dakota in the spring of 1995. (Written public comment was also provided by the citizens of California, where a forum was scheduled but, regrettably, cancelled due to coinciding activities of the U.S. Secretary of Agriculture.)

This volume, to be complemented by the committee's deliberative report, presents much of the data generated during the first stage of the process. It draws heavily on data and information already in the public domain that pertain to the colleges' activities. For example, the Current Research Information System (CRIS), administered and maintained by the U.S. Department of Agriculture (USDA), reports sources of funding for agricultural research activities and describes the allocation of those funds, institution by institution, among research goals, program areas, and commodities. The Food and Agricultural Education Information System (FAEIS), sponsored by the USDA and maintained at Texas A&M University, reports trends in enrollment, graduates, and degree programs at colleges of agriculture and related colleges and schools, and student and faculty demographics. Most of the CRIS and FAEIS data is self-reported by the colleges of agriculture and other reporting institutions. This publication reproduces data using categories developed by those information systems in collaboration with reporting institutions.

The committee recognizes that the data in this volume represent a starting point. Oftentimes the presentation of data raises at least as many questions as it answers. For example, the research data collected by CRIS and presented in chapters 4 and 7 offer a way to track the types of problems the college system is attempting to solve; alone the data do not allow an assessment of the quality of the system's research or its success at solving agricultural and food systems problems. (A summary of results of studies assessing the economic returns to agricultural research is presented in Chapter 2.)

The answers to questions of quality or successful problem solving require more research and analysis and pose methodological challenges not at all unique to the agricultural research system. The education data presented in chapter 3 are, likewise, useful for evaluating whether the colleges of agriculture are attracting more or fewer of the nation's students, including women and minorities, and which academic specializations offered are of most and least interest to their students. These data do not by themselves indicate the quality of the college of agriculture education, the nature of curricula changes, or the responsiveness of college course material and training to the needs of the students' future employers. These questions are of interest, but they require additional analyses outside the scope of this report. The data presented to describe cooperative extension are also limited. They aid the understanding of the allocation of extension resources to different program areas; however, measures of the quality or success of these delivery services and of how well the allocations of extension resources correspond to community demands would also be highly desirable.

The committee notes two additional caveats with respect to the use of the data. First, the data are sometimes presented in such a way as to emphasize the size differences among the individual colleges in the nationwide system. Size differences are at times emphasized to highlight the need for an understanding of the distributional implications of national policies and programs for agricultural research and education. Data on size differences are *not* offered as measures of quality differences in either research or education programs across large and small colleges.

Second, the data offer snapshots of the college activities and programs at particular points in time. These static pictures do not necessarily capture the dynamics of the colleges' attempts to meet medium-and long-term goals in response to changing student and constituent demands. The ability to adjust over time to those changing interests and needs is constrained by resource availability, including financial resources and the knowledge and skills of the faculty, scientists, and staff.

Chapter 1 is a review of the history of the land grant system, reflected in federal legislation beginning in 1862, and an overview of the system as a whole as it is today. The colleges were instituted to serve specific needs appropriate to the nation's character at that time. The initiation of the colleges of agriculture—both the "1862s" and the historically black "1890s"—reflected the nation's largely rural population and farm-economy base—and the racial separateness of the time. It is important to understand these roots. Since the system was designed to serve the public of yesterday, how is it adapting to changing times to serve the public of today and of the future?

Chapter 2 explores the colleges' operating environment. It reviews the characteristics of the U.S. economy and farming's role during the system's early years and goes on to illustrate how very different the U.S. economy, agriculture, and farming are today. First, and perhaps foremost, the United States is now a country of urbanites and suburbanites, few of whom retain ties to farming. Second, the majority of U.S. farm output is provided by only a small percent of all farms, and for the majority of the remaining farms farming receipts provide only a portion of the farm family's total income. Third, many of today's rural communities, although often less well-off economically than nonrural communities, have little or no economic base in farming.

The original colleges were mandated to serve the needs of the farmer, farm family, rural community, and national economy, which were closely intertwined in the system's early years. They were also mandated to do research because farms were too small to do their own. Indeed, if a farmer were to invest in developing new production technologies, the technologies could too quickly and easily be adopted by neighbors for the investment to pay off. Today's conditions are very different for some farmers. Today successful commercial farmers may oversee a huge corporate enterprise; interact productively with private seed, chemical, equipment, crop consulting, and biotechnology firms; conduct their own research or contract with private research firms; and have an array of sophisticated production and information technologies available. At the same time, a large share of the nation's smaller volume, part-time, limited-resource, or "hobby" farmers may share few or none of these traits. It is important to understand these changes. If today there are very few farmers in relation to urbanites, suburbanites, and nonfarm rural dwellers—and there are in fact many different types of farmers—how does today's college of agriculture define its constituency and shape a new public service role in a modern context?

Teaching, research, and extension are explored in chapters 3, 4, and 5, respectively. Each of these three functions has evolved somewhat differently in response to different administrators, constituent pressures, university rewards and incentives, and funding bases. Chapter 3 considers the role the colleges play in educating the nation's undergraduate and graduate students and how the changing demand for that education is reflected in enrollment, number of graduates, and number and category of degree programs. A comparison of the distribution of students across academic specializations with the distribution of research specializations might form the basis for discussing the congruence between the teaching and research programs or for suggesting changes in curriculum and faculty expertise. The data show that a relatively small number of the schools educate a disproportionately larger share of the students. Chapter 3 also reports demographic characteristics of an important category of college of agriculture graduates—Ph.D. agricultural scientists who often become college teachers, scientists, and administrators. The demographic characteristics of the colleges' clientele have changed dramatically over time. Not only are nonfarm constituents much more influential, but also communities in some parts of the country are much more ethnically diverse. Have demographic characteristics of college staff and graduates changed, too? How important might a more diverse set of characteristics be to addressing the needs and priorities of a more diverse public?

Research and extension functions are explored in a similar manner in chapters 4 and 5, respectively. Of particular interest in Chapter 4 is the allocation of research funds and staff to different types of research problems. Examining the distribution of research funds might reveal something of the types of problems the system is attempting to solve or the needs it is attempting to address. Although data on research expenditures or scientist years are measures of research *input,* rather than *output* or *benefits*, they might be useful first approximations of what types of interests or goals are primarily served by the system's research. Understanding who benefits from research, for example, could lead to new proposals for who should pay the colleges' research costs. In addition, a comparison of those allocations today with allocations in the past may be a useful indicator of the colleges' progress in adapting to the demands of a changing clientele. Comparisons of how research scientists and extension staff use their time—found in Chapter 5—may also lead to interesting questions: What do the results of these comparisons suggest about the future of the research-extension interface that has characterized the colleges' traditional work?

Chapter 6 explores the system's components from a different perspective—that of the federal-state-private sector partnerships that jointly support the system's three functions. The nature of the colleges' continuing ties to the federal government, through the U.S. Department of Agriculture (USDA), are clearly unique among the nation's science and research institutions. These ties include federal funds for research and extension allocated to each state using a formula that has changed little if at all over the decades and a federal requirement for state matching grants, which is what drew the states into the partnership. If legislators were proposing a formula-based funding mechanism today, against the context of today's state economies, how might it differ from the one proposed many decades ago? Despite the continuing federal presence in the system, that role is significantly smaller today than in the past, and the roles of state governments and the private sector are larger. The data in this chapter provide a starting point for asking questions about the future role of the federal government in the system and what its purpose or goals might be.

Chapter 7 offers yet another perspective on the system's components by providing a closer look at individual colleges and at their similarities and differences within the system as a whole. The pressures weighing on the colleges of agriculture are not borne equally by the many separate institutions. Across the country, states differ significantly in the prominence of their rural versus urban populations and the demographics of their communities; the characteristics and importance of their farm, food, fiber, and natural resource sectors; the makeup of interest groups and political forces; and the traits of their higher-education systems. Thus each college has a somewhat different constituency and faces a somewhat different set of pressures for change. In their earlier years, colleges were probably more similar than they are today; and yet today they are probably still more similar than their distinct environments would suggest. The data in this chapter could provide a first step toward assessing whether the colleges are independently and differently adapting to change. The data may serve as the catalyst to thinking about how the system might evolve to include a set of institutions more specialized in terms of functional orientation (what they do); emphasis on teaching, research, and extension programs (in whose interest they do it); and the nature of their public and private partnerships (who supports what they do). In addition, the data seem to identify pronounced differences across institutions—for example, the size of research expenditures. These differences may initiate questions about the future of smaller institutions and whether there is a potential role for federal policy in balancing the inherent advantages of larger ones.

In sum, this publication represents the culmination of the first stage of the committee's study of the colleges of agriculture. The deliberative report composed of conclusions and recommendations for institutional change and public policy will follow. The data in the report, compiled from many published or publicly available sources, and adhering to the categorization of data established by those sources, were collected to facilitate the committee's deliberations. These data might not provide clear answers, but they do help in formulating and exploring well-founded questions. The data are being published to contribute to the public's understanding of this distinctly public system and to the public debate regarding its future.

Anthony S. Earl, *Chair*

Committee on the Future of the Colleges of Agriculture in the LandGrant University System

# ACKNOWLEDGEMENTS

This publication was compiled with help from many people who provided data, assisted with data base searches, and suggested literature sources. We especially thank Sherry Stebenne Whately, project coordinator, USDA Food and Agricultural Education Information System (FAEIS) and John Myers, director, and Dennis Unglesbee of the USDA Current Research Information System (CRIS).

# CONTENTS

# TABLES AND FIGURES

## TABLES

**Appendix Tables**



## FIGURES

# COLLEGES OF AGRICULTURE AT THE LAND GRANT UNIVERSITIES

# 1

# HISTORY AND OVERVIEW OF THE LAND GRANT COLLEGE SYSTEM

*This chapter reviews the legislative origins of today's land grant university system, including the federal mandates to provide instruction in agriculture and the mechanical arts, conduct agricultural research, and deliver knowledge and practical information to farmers and consumers. The chapter also describes the geographical dimension of the system's infrastructure by providing names and locations of land grant colleges of agriculture and of the related colleges and schools of forestry and veterinary medicine.*

## WHAT ARE LAND GRANT COLLEGES OF AGRICULTURE?

- The history of land grant colleges of agriculture is intertwined with the history of higher education for U.S. citizens of average means. The land grant system began in 1862 with a piece of legislation known as the Morrill Act (see box copy, p. 2). This law gave states public lands provided the lands be sold or used for profit and the proceeds used to establish at least one college—hence, land grant colleges—that would teach agriculture and the mechanical arts. Land grants for the establishment of colleges of agriculture and mechanical arts were also later given to U.S. territories and the District of Columbia. The legislative mandate for these land grant colleges helped extend higher education to broad segments of the U.S. population.
- Public universities existed already in some states; however, most states responded to the Morrill Act by legislating new agricultural and mechanical arts colleges rather than by endowing existing state institutions (Kerr, 1987). The act gave rise to a network of often poorly financed colleges known as the "1862s" (Table 1-1; Figure 1-1). The Second Morrill Act, which provided for annual appropriations to each state to support its land grant college, was passed by Congress in 1890.
- In addition to appropriating funding, the Second Morrill Act also forbade racial discrimination in admissions policies for colleges receiving these federal funds. A state could escape this provision, however, if separate institutions were maintained and the funds divided in a "just," but not necessarily equal, manner. Thus the 1890 act led to the establishment of land grant institutions for African Americans. Today there are 17 1890 institutions—including one private institution, Tuskegee University —located primarily in the southeast (Table 1-1; Figure 1-1). In addition to being part of the land grant system, these 17 1890 schools are among the more than 100 historically black colleges and universities in the United States.

**The First Morrill Act (1862): Donating Public Lands for Colleges of Agriculture and Mechanic Arts**

Section 4 (original). *And be it further enacted*, That all moneys derived from the sale of the lands aforesaid by the States to which the lands are apportioned, and from the sale of land scrip herein before provided for, shall be invested in stock of the United States, or of the States, or some other safe stocks, yielding not less than five per centum upon the par value of said stocks; and that the moneys so invested shall constitute a perpetual fund, the capital of which shall remain forever undiminished (except so far as may be provided in section fifth of this act), and the interest of which shall be inviolably appropriated, by each State which may take and claim the benefit of this act, to the endowment, support, and maintenance of at least one college where the leading object shall be, without excluding other scientific and classical studies, and including military tactics, to teach such branches of learning as are related to agriculture and the mechanic arts, in such manner as the legislatures of the States may respectively prescribe, in order to promote the liberal and practical education of the industrial classes in the several pursuits and professions in life.

- Over the decades, as the U.S. economy grew and changed, so did the nature of demands for education and scientific pursuit. As more and more U.S. citizens began to attend college, most colleges of agriculture were transformed into full-fledged universities. In some states, like California, Maryland, Minnesota, and Wisconsin, land grant universities have become the foremost public institutions of higher education and scientific research. In others, such as North Carolina, Michigan, and Oregon, higher education and research functions are shared with other prominent public institutions.
- Today, although many land grant universities are still known for their agricultural college roots, others have little agricultural identity and students are rarely from farm families. Despite their expansion well beyond the teaching of agriculture and mechanical arts, almost every land grant university still has a "college of agriculture"—colleges more similar to each other than are the universities where they are located.
- Over time, colleges of agriculture have been established at non-land grant institutions as well. The relative role of the non-land grants in educating students in agriculture-related academic specializations is discussed in Chapter 3.

A series of legislative acts endowed the colleges with a three-part function encompassing teaching, research, and extension.

**TABLE 1-1** Locations and Names of 1862 and 1890 Land Grant Colleges of Agriculture and Related Colleges and Schools of Forestry and Veterinary Medicine

| | | | | Names of Agriculture-Related Colleges | | |
|---|---|---|---|---|---|---|
| Location | Institution | Classification | Established | Agriculture | Forestry | Veterinary Medicine |
| Alabama | | | | | | |
| Auburn | Auburn U. | 1862 | 1856 | College of Agriculture | School of Forestry | College of Veterinary Medicine |
| Normal | Alabama A&M U. | 1890 | 1873 | School of Agriculture and Home Economics | None | None |
| Tuskegee | Tuskegee U. | 1890 | 1881 | College of Agriculture and Home Economics | None | School of Veterinary Medicine |
| Alaska | | | | | | |
| Fairbanks | U. of Alaska | 1862 | 1917 | School of Agriculture and Land Resources Management | [a] | None |
| American Samoa | | | | | | |
| Pago Pago | American Samoa Community College | 1862 | 1970 | Agricultural Program | None | None |
| Arizona | | | | | | |
| Flagstaff | Northern Arizona U. | NLG | 1899 | None | School of Forestry | None |
| Tucson | U. of Arizona | 1862 | 1885 | College of Agriculture | School of Renewable Natural Resources | None [b] |
| Arkansas | | | | | | |
| Fayetteville | U. of Arkansas | 1862 | 1871 | Division of Agriculture | [a] | None |
| Pine Bluff | U. of Arkansas | 1890 | 1873 | School of Agriculture and Home Economics | None | None |
| California | | | | | | |
| Arcata | Humboldt State U. | NLG | 1913 | None | College of Natural Resources | None |
| Berkeley | U. of California | 1862 | 1868 | College of Natural Resources | [a] | None |
| Davis | U. of California | 1862 | 1908 | College of Agricultural and Environmental Sciences | None | School of Veterinary Medicine |
| Riverside | U. of California | 1862 | 1907 | College of Natural and Agricultural Sciences | None | None |
| San Luis Obispo | California Polytechnic State U. | NLG | 1901 | College of Agriculture | [a] | None |
| Colorado | | | | | | |
| Fort Collins | Colorado State U. | 1862 | 1870 | College of Agricultural Sciences | College of Natural Resources Sciences | College of Veterinary Medicine and Biomedical Sciences |
| Connecticut | | | | | | |
| New Haven | Yale U. | NLG | 1701 | None | School of Forestry and Natural Resources | None |
| Storrs | U. of Connecticut | 1862 | 1881 | College of Agriculture and Natural Resources | [a] | None |
| Delaware | | | | | | |
| Dover | Delaware State U. | 1890 | 1891 | Department of Agriculture and Natural Resources | None | None |
| Newark | U. of Delaware | 1862 | 1743 | College of Agricultural Sciences | [a] | None |

**TABLE 1-1**

| Location | Institution | Classification | Established | Names of Agriculture-Related Colleges: Agriculture | Forestry | Veterinary Medicine |
|---|---|---|---|---|---|---|
| District of Columbia | | | | | | |
| Washington, D.C. | U. of the District of Columbia | 1862 | 1851 | College of Life Sciences | None | None |
| Florida | | | | | | |
| Gainesville | U. of Florida | 1862 | 1853 | College of Agriculture and Conservation | School of Forest Resources | College of Veterinary Medicine |
| Tallahassee | Florida A&M U. | 1890 | 1887 | College of Engineering, Science, and Technology | None | None |
| Georgia | | | | | | |
| Athens | U. of Georgia | 1862 | 1785 | College of Agriculture and Environmental Sciences | School of Forest Resources | College of Veterinary Medicine |
| Fort Valley | The Fort Valley State College | 1890 | 1895 | School of Agriculture, Home Economics, and Applied Programs | None | [b] |
| Guam | | | | | | |
| Mangilao | U. of Guam | 1862 | 1952 | College of Agriculture and Life Sciences | [a] | None |
| Hawaii | | | | | | |
| Honolulu | U. of Hawaii | 1862 | 1907 | College of Tropical Agriculture and Human Resources | [a] | None |
| Idaho | | | | | | |
| Moscow | U. of Idaho | 1862 | 1889 | College of Agriculture | College of Forestry, Wildlife, and Range Sciences | [b] |
| Illinois | | | | | | |
| Carbondale | Southern Illinois U. | NLG | 1874 | College of Agriculture | Department of Forestry | None |
| Urbana | U. of Illinois | 1862 | 1867 | College of Agriculture | [a] | College of Veterinary Medicine |
| Indiana | | | | | | |
| West Lafayette | Purdue U. | 1862 | 1869 | College of Agriculture | [a] | School of Veterinary Medicine |
| Iowa | | | | | | |
| Ames | Iowa State U. of Science and Technology | 1862 | 1858 | College of Agriculture | [a] | College of Veterinary Medicine |
| Kansas | | | | | | |
| Manhattan | Kansas State U. | 1862 | 1863 | College of Agriculture | [a] | College of Veterinary Medicine |
| Kentucky | | | | | | |
| Frankfort | Kentucky State U. | 1890 | 1886 | | None | None |
| Lexington | U. of Kentucky | 1862 | 1866 | College of Agriculture | [a] | [b] |
| Louisiana | | | | | | |
| Baton Rouge | Louisiana State U | 1862 | 1860 | College of Agriculture | [a] | College of Veterinary Medicine |
| Baton Rouge | Southern U. and A&M College | 1890 | 1880 | College of Agriculture and Home Economics | None | None |
| Ruston | Louisiana Tech U. | NLG | 1894 | None | School of Forestry, Wildlife, and Fisheries | None |

| | | | | | | |
|---|---|---|---|---|---|---|
| Maine | | | | | | |
| Orono | U. of Maine | 1862 | 1865 | College of Natural Resources, Forestry, and Agriculture | [a] | [b] |
| Maryland | | | | | | |
| College Park | U. of Maryland | 1862 | 1856 | College of Agriculture | [a] | None |
| Princess Anne | U. of Maryland-Eastern Shore | 1890 | 1886 | School of Agricultural Sciences | None | None |
| Massachusetts | | | | | | |
| Amherst | U. of Massachusetts | 1862 | 1821 | College of Food and Natural Resources | [a] | [b] |
| Medford | Tufts U. | NLG | 1852 | None | None | School of Veterinary Medicine |
| Michigan | | | | | | |
| Ann Arbor | U. of Michigan | NLG | 1817 | None | School of Natural Resources | None |
| East Lansing | Michigan State U. | 1862 | 1855 | College of Agriculture and Natural Resources | [a] | College of Veterinary Medicine |
| Houghton | Michigan Technological U. | NLG | 1885 | None | School of Forestry and Wood Products | None |
| Micronesia | | | | | | |
| Kolonia | College of Micronesia | 1862 | | College of Tropical Agriculture and Sciences | None | None |
| Minnesota | | | | | | |
| Minneapolis | U. of Minnesota | 1862 | 1851 | College of Agricultural, Food, and Environmental Sciences | College of Natural Resources | College of Veterinary Medicine |
| Mississippi | | | | | | |
| Lorman | Alcorn State U. | 1890 | 1871 | Division of Agricultural Research and Applied Economics | None | None |
| State College | Mississippi State U. | 1862 | 1878 | College of Agriculture and Home Economics | School of Forest Resources | College of Veterinary Medicine |
| Missouri | | | | | | |
| Columbia | U. of Missouri | 1862 | 1839 | College of Agriculture, Food, and Natural Resources | [a] | College of Veterinary Medicine |
| Jefferson City | Lincoln U. | 1890 | 1866 | College of Agriculture, Applied Sciences, and Technology | None | None |
| Montana | | | | | | |
| Bozeman | Montana State U. | 1862 | 1893 | College of Agriculture | [a] | [b] |
| Missoula | U. of Montana | NLG | 1893 | None | School of Forestry | None |
| Nebraska | | | | | | |
| Lincoln | U. of Nebraska | 1862 | 1869 | College of Agricultural Sciences and Natural Resources | [a] | [b] |
| Nevada | | | | | | |
| Reno | U. of Nevada | 1862 | 1874 | College of Agriculture | [a] | [b] |
| New Hampshire | | | | | | |
| Durham | U. of New Hampshire | 1862 | 1866 | College of Life Sciences and Agriculture | [a] | None |
| New Jersey | | | | | | |
| New Brunswick | Rutgers-The State U. | 1862 | 1766 | Cook College | [a] | None |
| New Mexico | | | | | | |
| Las Cruces | New Mexico State U. | 1862 | 1888 | College of Agriculture and Home Economics | [a] | None |

**TABLE 1-1**

| Location | Institution | Classification | Established | Names of Agriculture-Related Colleges: Agriculture | Forestry | Veterinary Medicine |
|---|---|---|---|---|---|---|
| New York | | | | | | |
| Ithaca | Cornell U. | 1862 | 1865 | College of Agriculture and Life Sciences | [a] | New York Veterinary College |
| Syracuse | State U. of New York | NLG | 1911 | None | College of Environmental Science and Forestry | None |
| North Carolina | | | | | | |
| Greensboro | North Carolina A&T State U. | 1890 | 1891 | School of Agriculture | None | None |
| Durham | Duke U. | NLG | 1838 | None | School of the Environment | |
| Raleigh | North Carolina State U. | 1862 | 1887 | College of Agriculture and Life Sciences | College of Forest Resources | School of Veterinary Medicine |
| North Dakota | | | | | | |
| Fargo | North Dakota State U. | 1862 | 1890 | College of Agriculture | [a] | None |
| Northern Marianas | | | | | | |
| Saipan | Northern Marianas College | 1862 | | School of Agriculture and Life Sciences | None | None |
| Ohio | | | | | | |
| Columbus | The Ohio State U. | 1862 | 1870 | College of Food, Agriculture, and Environmental Sciences | None | College of Veterinary Medicine |
| Oklahoma | | | | | | |
| Langston | Langston U. | 1890 | 1897 | Department of Agriculture | None | None |
| Stillwater | Oklahoma State U. | 1862 | 1890 | Division of Agricultural Sciences and Natural Resources | [a] | College of Veterinary Medicine |
| Oregon | | | | | | |
| Corvallis | Oregon State U. | 1862 | 1868 | College of Agricultural Sciences | College of Forestry | College of Veterinary Medicine |
| Pennsylvania | | | | | | |
| Philadelphia | U. of Pennsylvania | NLG | 1740 | None | None | School of Veterinary Medicine |
| University Park | The Pennsylvania State U. | 1862 | 1855 | College of Agricultural Sciences | School of Forest Resources | [b] |
| Puerto Rico | | | | | | |
| Rio Piedras | U. of Puerto Rico | 1862 | 1900 | College of Agricultural Sciences | [a] | None |
| Rhode Island | | | | | | |
| Kingston | U. of Rhode Island | 1862 | 1892 | College of Resource Development | [a] | [b] |
| South Carolina | | | | | | |
| Clemson | Clemson U. | 1862 | 1889 | College of Agricultural Science | College of Forest and Recreational Resources | [b] |
| Orangeburg | South Carolina State College | 1890 | 1896 | Department of Agribusiness and Economics | None | None |
| South Dakota | | | | | | |
| Brookings | South Dakota State U. | 1862 | 1884 | College of Agriculture and Biological Sciences | [a] | [b] |

| | | | | | | |
|---|---|---|---|---|---|---|
| Tennessee | | | | | | |
| Knoxville | U. of Tennessee | 1862 | 1794 | College of Agricultural Sciences and Natural Resources | [a] | College of Veterinary Medicine |
| Nashville | Tennessee State U. | 1890 | 1912 | School of Agriculture and Home Economics | None | None |
| Texas | | | | | | |
| College Station | Texas A&M U. | 1862 | 1876 | College of Agriculture and Life Sciences | [a] | College of Veterinary Medicine |
| Nacogdoches | Stephen F. Austin State U. | NLG | 1917 | None | School of Forestry | None |
| Prairie View | Prairie View A&M U. | 1890 | 1876 | Department of Agriculture | None | None |
| Utah | | | | | | |
| Logan | Utah State U. | 1862 | 1888 | College of Agriculture | [a] | [b] |
| Vermont | | | | | | |
| Burlington | U. of Vermont | 1862 | 1791 | College of Agriculture and Life Sciences | School of Natural Resources | None |
| Virgin Islands | | | | | | |
| St. Thomas | U. of the Virgin Islands | 1862 | 1963 | College Administration | [a] | None |
| Virginia | | | | | | |
| Blacksburg | Virginia Polytechnic Institute and State U. | 1862 | 1872 | College of Agriculture and Life Sciences | School of Forestry and Wildlife Sciences | Virginia-Maryland Regional College of Veterinary Medicine |
| Petersburg | Virginia State U. | 1890 | 1882 | School of Agriculture and Applied Sciences | None | None |
| Washington | | | | | | |
| Pullman | Washington State U. | 1862 | 1890 | College of Agriculture and Home Economics | [a] | College of Veterinary Medicine |
| Seattle | U. of Washington | NLG | 1861 | None | College of Forest Resources | None |
| West Virginia | | | | | | |
| Morgantown | West Virginia U. | 1862 | 1867 | College of Agriculture and Forestry | Division of Forestry | None |
| Wisconsin | | | | | | |
| Madison | U. of Wisconsin | 1862 | 1836 | College of Agriculture and Life Sciences | School of Natural Resources | School of Veterinary Medicine |
| Wyoming | | | | | | |
| Laramie | U. of Wyoming | 1862 | 1886 | College of Agriculture | [a] | [b] |

NOTE: Non-land grant (NLG) institutions included in this table have separate colleges or schools of forestry and/or veterinary medicine or a college of agriculture that has a forestry or veterinary medicine program. Information about other non-land grant colleges of agriculture may be obtained from the American Association of State Colleges of Agriculture and Renewable Resources (AASCARR). Information provided here was the most recent available at the time the table was compiled.

[a]The college of agriculture has a forestry program.
[b]The college of agriculture has a veterinary science program.

SOURCES: Data were derived from USDA data bases—Food and Agricultural Education Information System (FAEIS) and Current Research Information System (CRIS)— and from cooperating state institutions and from the U.S. Department of Agriculture. 1993–1994. Directory of Professional Workers in State Agricultural Experiment Stations.

**FIGURE 1-1**
*Map shows locations of the 1862 and 1890 land grant colleges and universities in the contiguous United States, Alaska, and Hawaii. Not shown are land grant locations at American Samoa, Guam, Micronesia, Northern Marianas, Puerto Rico, and St. Thomas in the U.S. Virgin Islands. Symbol placement indicates geographic location of each institution, showing physical proximity.*

- The 1862 Morrill Act gave the land grant colleges their mandate to teach. The colleges acquired a research function in 1887 through the Hatch Act, which recognized the need for original research to underpin the teaching of agriculture and help develop agricultural innovations. The legislation funded a system of state agricultural experiment stations (SAESs), most of which were established under the direction of the 1862 land grant colleges. Table 1-2 outlines a chronological progression of legislation mandating the many iterations of the land grant college system.
- Today SAESs operate in conjunction with and, in almost all cases, on locations at colleges of agriculture. Connecticut and New York, in addition to on-campus SAESs, have an off-campus SAES. Many other states have branch stations, that is, SAES subsidiaries located off campus and often in agricultural areas of direct interest to the branch station's research.
- Most faculty at land grant colleges of agriculture have SAES appointments. This grants them potential access to "Hatch" research funds, which are administered by USDA and funneled to the SAESs on a formula basis. Some faculty scientists who have SAES appointments also conduct research at other colleges that have related programs, such as in the life sciences. The SAES director and the dean of the college of agriculture are usually, but not always, the same person.
- With the 1914 Smith-Lever Act, the colleges took on a third function, called "extension," which was designed to disseminate agricultural college-generated knowledge beyond the campus to farms and consumers. Extension was to be a cooperative activity between the federal government (through USDA) and the states (through the land grant colleges). County governments, through a network of county extension agents, soon became cooperative extension partners.

**TABLE 1-2** Chronology of Major Legislation Affecting the Land Grant Colleges of Agriculture System

| Year | Legislation | Provisions | Key Result | Funding Mechanism |
|---|---|---|---|---|
| 1862 | Act of Congress | Mandated the U.S. Department of Agriculture (USDA) to take over agricultural science functions of the Patent Office. | Established the office of Commissioner of Agriculture. | |
| 1862 | Morrill Act | Provided for land on which each state could establish and maintain at least one college to teach (without excluding other scientific and classical studies and including military tactics) courses related to agriculture and mechanical arts in order to promote the liberal and practical education of the industrial classes. | Established the land grant college system. | Each state would receive 30,000 acres of land for each senator and representative in Congress. States where not enough public land was available were given scrip to public land in other states; the income from the land to be used for operating expenses (construction, purchase, repair of buildings excluded). |
| 1887 | Hatch Act | Sanctioned each state to establish an experiment station to conduct original research or verify experiments bearing directly on the agricultural industry of the United States. Stations were to be established under direction of land grant colleges, but exceptions were permitted. | Established the state agricultural experiment stations (SAESs). | Each qualifying state would receive $15,000 per year. |
| 1890 | Second Morrill Act | First proposed in 1872, act provided for direct annual appropriations to each state to further support land grant colleges.<br>Each state could receive additional funds to more completely endow and support land grant colleges. The funds were to pay for instruction in agriculture, mechanical arts, the English language, and branches of mathematics, physical, natural, and economic sciences related to agriculture and mechanical arts. African Americans were to be admitted to land grant institutions. States could establish separate land grant colleges for African Americans. | Forbade racial discrimination in admission to colleges receiving funds and gave rise to the so called "1890 colleges." | Congress would give each qualifying state $15,000 the first year and increase the amount by $1,000 per year for subsequent years until the annual amount reached $25,000. |
| 1906 | Adams Act | Provided each state additional federal funding to pay the necessary expenses of conducting original research and experiments. | Emphasized science, and more accountability coincided with the formation of Experiment Station Committee on Organization and Policy (ESCOP). | Each qualifying state would receive a maximum of an additional $15,000 per year. Each state was entitled to an increase of $5,000 for the first year and $2,000 more than the previous year's sum for 5 subsequent years. |
| 1907 | Nelson Amendment | Same as Second Morrill Act with the additional specification that a portion of the fund could be used for "providing courses for the special preparation of instructors for teaching the elements of agriculture and mechanic arts." | | Doubled annual appropriation to $50,000. |

**TABLE 1-2**

| Year | Legislation | Provisions | Key Result | Funding Mechanism |
|---|---|---|---|---|
| 1914 | Smith-Lever Act | Created Cooperative Extension Service to aid in disseminating to the public useful and practical information about subjects relating to agriculture and home economics and to encourage its application.<br>The land grant colleges and USDA were to cooperate in extension work, which was to consist of instruction and practical demonstrations in agriculture and home economics to persons not attending the land grant college. Information was to be supplied through field demonstrations. | Established the Cooperative Extension Service. | Provided lump sum of $10,000 per state ($480,000 total) and additional formula funding. Formula funds were based on what percentage of the total U.S. rural population resided in the state. Formula funding phased in over 7 years, to a maximum of $4.1 million. The formula money was to be matched by state funds. |
| 1917 | Smith-Hughes Vocational Education Act | Made federal grants available to eligible states to stimulate vocational education in agriculture, home economics, and industrial arts. Grants were for (a) training of teachers by public colleges and (b) funding part of the salaries of teachers and directors of vocational agricultural subjects in secondary public schools. | | No specific statement made regarding funding. |
| 1925 | Purnell Act | Each state could receive additional federal funding for research to (a) establish and maintain a permanent and efficient agricultural industry and (b) develop and improve the rural home and rural life. | Provided first emphasis on economics, home economics, and sociology. | Each qualifying state could receive a maximum of $30,000 per year. Each state was entitled to an increase of $10,000 for the first year and $5,000 over the previous year's sum for 4 subsequent years. |
| 1928 | Capper-Ketcham Act | | Provided for expansion of Cooperative Extension Service | An additional lump sum grant of $20,000 per state ($980,000 total per year) and an additional $500,000 starting in 1929 to be allocated by formula.<br>Required 1/3 of added funds to be matched in 1923 and full matching after 1928. |
| 1935 | Bankhead-Jones Act | Research: SAESs and USDA could receive additional funding for research into basic problems of agriculture; research relating to quality improvement, new and improved methods of production and distribution, and new and extended uses and markets for agricultural commodities; and research relating to conservation, development, and recreational use of land and water.<br><br>Extension: Provided for expansion of Cooperative Extension Service | Established formula funding for research and federal-state matching grants. | Research: A maximum of $5 million per year, with $3 million to the SAESs. A total increment of $1 million per year for each of 5 years. Funds to be distributed to the states on the basis of what percentage of total U.S. rural population resided in their state, and each state must match federal contribution with nonfederal funding of the SAES.<br><br>Extension: An additional lump sum grant of $20,000 per state ($980,000 total per year) and an additional $8 million allocated to states by formula in 1936 and $1 million additional in each of the next 4 years. Formula funds to be allocated by state's share of the U.S. farm population; matching not required. |

| | | | | |
|---|---|---|---|---|
| 1945 | Bankhead-Flannagan Act | | Provided for further expansion of Extension Service. | Two percent of the federal appropriation was for federal administration, 4% was set aside for the Secretary for special need allocation, and 94% distributed by a formula based on a state's percentage of the total U.S. farm population. |
| 1946 | Research and Marketing Act | SAESs and USDA could receive additional funding for marketing and utilization research and for regional research involving two or more states involved in finding a solution to a problem of regional significance. | Introduced open-ended appropriations. Linked agricultural research and development to national welfare. Stated goals of Congress to maintain a balanced farming and industrial economy. Established farming and industrial national advisory committee. | Title I, Section 9: Total SAES funding up by $2.5 million in 1947 and 1948; $5 million increase for each of 1949, 1950, and 1951; such additional funds as Congress shall deem necessary for additional years. Allocation among states: 20% equally among states; 26% by formula according to state's percentage of U.S. rural population; 26% by formula according to a state's percentage of total U.S. farm population; 25% for regional research; 3% for federal administration.<br><br>Title I, Section 10: Increased USDA funding for research. Authorized grants for "new uses" research to increase from $3 million in 1947 to $15 million after 1950; funds for cooperative research into farm product utilization to rise from $1.5 million in 1947 to $6 million after 1950; reauthorized $2 million annual Special Research Fund provided for in the Bankhead-Jones Act of 1935.<br><br>Title II: Authorized an additional $2.5 million in 1947 and increasing to $20 million per year after 1950 for marketing research, carried out cooperatively with SAESs and other public and private institutions, on a matching grant basis. |
| 1953 | Amended Smith-Lever Act | Consolidated nine existing acts and provided for appropriations for federal extension staff in USDA. | | Provided that subsequent increases be allocated 4% to special need; 48% based on a state's percentage of the U.S. farm population; and 48% based on a state's share of the rural population. |
| 1955 | Amended Hatch Act | Proposed to support research contributing to the maintenance of a permanent and effective agricultural industry in the United States, including research basic to the problems of agriculture in its broadest aspects and research related to the development and improvement of the rural home and rural life and the maximum contribution of agriculture to the welfare of the consumer. Removed restrictions on buildings, but Hatch funds still had to be spent within the year awarded. | Retained allocation formulas, matching-grant requirements, and "open-ended" appropriations. Congress rejected a proposal to reduce marketing research by 20% and insisted that earmarking apply to all increases in appropriations. | Consolidated federal funding for SAESs into two accounts (formula funds and regional research funds). No set annual amounts were established. Allocation was according to the formula from the 1946 Research and Marketing Act: 20% of each year's appropriation equally among states; 26% by formula according to a state's percentage of the U.S rural population; 26% by formula according to a state's share of the U.S. farm population; 25% for cooperative regional research; 3% for federal administration. |
| 1955 | Smith-Lever Amendment | Provided for establishment of Special Program system. | | Provisions added permitted special nonformula funds. |

*continues*

**TABLE 1-2**

| Year | Legislation | Provisions | Key Result | Funding Mechanism |
|---|---|---|---|---|
| 1960 | Amendment to Title II, Section 22 of the Bankhead-Jones Act Stennis Act | Same as Morrill Act of 1862 as "amended and supplemented." | | Annual appropriation of $7,650,000 distributed equally among the states and Puerto Rico; $4,300,000 allotted based on the proportion of state (Puerto Rico) population to total U.S. and Puerto Rico population. |
| 1961 | Amended Smith-Lever Act | Resource and community development extension added. | | Provided $700,000 per year for resource and community development work. |
| 1962 | Smith-Lever Amendment | | | Froze distribution of current federal funds to each state. Subsequent increases to be 4% to the federal service and, of the remainder, 20% in equal proportions to all states and 40% based on a state's percentage of the U.S. rural population and 40% according to its percentage of the U.S. farm population. |
| 1962 | McIntire-Stennis Forestry Research Act | Made funding available to SAESs, land grant colleges, and forestry schools for forestry research—including reforestation, woodlands and related watershed management, outdoor recreation, wildlife habitats, wood utilization, and such other studies as may be necessary to obtain the fullest and most effective use of forest resources. | Coincided with the formation of the Cooperative State Research Service (CSRS) in 1961–1963. CSRS to administer appropriations under McIntire-Stennis Act. | A formula allocated $10,000 to each state, 40% of the remainder according to a state's share of the nation's total commercial forest land, 40% according to the value of its timber cut annually, and 20% according to its contribution of nonfederal forestry research dollars. In both 1964 and 1965, $1 million was appropriated, 2% of CSRS-managed funds (by 1974 this figure had increased to more than $6 million annually, 7% of CSRS-managed money, and by 1984 it was up to almost $13 million or 6% of combined federal funding to the states for agricultural research and development). |
| 1965 | Research Facilities Act | Earmarked funds to be matched by the states for the construction, acquisition, and remodeling of buildings, laboratories, and other capital facilities. Supported new construction only of facilities for research on hazardous chemicals used in farming. Allowed each station to obligate its annual share over a 3-year period for the first time. | | The formula resembled that of the amended Hatch Act: one-third equally to each state; one-third based on the proportion of rural residents; one-third based on the proportion of farm population. Total allocations were $3.2 million in 1965, $2 million per year in 1966, 1967, and 1968; none was provided in 1969; and $1 million in 1970, for the last time. |
| 1965 | Public Law 89-106 | Established "Specific Research Grants" program to finance selected projects over a maximum of 5 years. Later became the "Special Grants" program. Earmarked funds to address specific problems of constituent concern or multistate problems. | | CSRS would call annually for proposals in areas singled out by Congress for special attention. In 1966, $1.6 million was offered; in 1967–1970, $1.7 million per year. $283,000 per year was allocated to the 16 1890 institutions, for an average of $17,658 each. In 1972 agricultural research allocation was increased significantly, and Tuskeegee University became eligible to receive these funds, making it the seventeenth 1890 institution. |

| | | | | |
|---|---|---|---|---|
| 1968 | | | | Congress abolished special program funding except for $1.6 million for agricultural marketing. These funds were to be allocated by formula. |
| 1972 | Federal Rural Development Act | Research: SAESs and Extension Service could receive funds for rural development and small-farm research and extension.<br><br>Extension: Title V authorized work in rural communities in agriculture and nonagriculture fields. | | The 1972 Act authorized $10 million for 1974, $15 million for 1975, and $20 million for 1976. Actual expenditures were much less. $3 million was provided in each of the first 3 years, split between extension and research, and allocated among the SAESs on a basis similar to the Hatch formula, except that 10% was reserved for interstate projects. Funding continued at $3 million per year for another 4 years after the initial authorization expired in 1977.<br>Funds were to be distributed 4% for federal administration, 10% for multistate work, 20% equally distributed among states, and 33% each according to a state's percentage of the U.S. rural and farm population. |
| 1977 | National Agricultural Research, Extension and Teaching Policy Act (Title XIV of the Food and Agriculture Act of 1977) | Continued and strengthened amended Hatch programs and initiated a new competitive grants program for high-priority research, open to all scientists, to be awarded on a competitive basis to private- and public-sector organizations, including SAESs, all colleges and universities, other research organizations, federal agencies, and individuals. Continued the Special Grants program. Dropped the requirement that 20% of amended Hatch funds be earmarked for marketing research.<br>Transferred administration of the Bankhead-Jones Act from Office of Education to USDA. Provided formula funds for research at 1890 institutions. | Other new earmarked grants also introduced (e.g., energy research; and animal health). New mechanisms for more formalized research planning, central (federal) direction, and accountability. | Hatch formula funds were strengthened with $120 million called for in 1978 and increases of $25 million per year up to $220 million in 1982. Allocation was basically as by previous arrangements and formulas. The competitive grants program authorized additional spending of $25 million/year in 1978, $30 million in 1979, $35 million in 1980, $40 million in 1981, and $50 million in 1982.<br>Permanent or sustained institutional federal funding via Section 1445 of the act—the Evans-Allen Research Program—provided formula-funded programs for 1890 institutions.<br>The Rural Development Title V formula of 1972 was changed to 19% for farm research programs and to 77% for small-farm extension programs. |
| 1978 | Passage of the Resource Extension Act | Authorized funding for extension programs in forestry and other renewable national resources. | | Funding is by Congressional appropriation. |
| 1981 | Amendments to Title IV (National Agricultural Research, Extension and Teaching Policy Act of 1977) | Primarily extended the 1977 act for 4 years. Introduced $10 million annual rangeland research program and $7.5 million annual aquaculture research program.<br>Rural development extension funds became part of Smith-Lever formula appropriation. | Congress effectively promised not to replace, but to supplement, formula funds with competitive grants. | (See 1977 act.) Hatch funds were authorized to increase from $220 million in 1982 to $250 million in 1985. Hatch funds were guaranteed at a minimum of 25% of USDA expenditures in cooperative programs. |

**TABLE 1-2**

| Year | Legislation | Provisions | Key Result | Funding Mechanism |
|---|---|---|---|---|
| 1985 | National Agricultural Research, Extension and Teaching Policy Act (Title IV of the Food Security Act of 1981) | Primarily extended the 1981 act for 4 years. Added a new subtitle to promote sustainable agriculture. Earmarked funds for marketing research were reintroduced ($10 million per year) along with Trade Development Centers at land grant universities (on a matching basis). | USDA permitted to fund competitive grants for facilities at SAESs. | (See 1981 act.) Hatch funds were to increase only 4% per year, while competitive grants were to increase substantially, especially for biotechnology research. Hatch funding of $270 million in 1986 to increase to $310 million for 1990. Competitive Grants funding to increase from $50 million in 1985 to $70 million in 1986 and subsequent years.<br>Amended the 1977 act to provide not less than 6% of Smith-Lever funds be allocated for extension work at the 1890 institutions. |
| 1990 | Food, Agriculture, Conservation, and Trade Act (farm bill) | Reauthorized sustainable agriculture research and education program and added new program for training of extension service personnel in sustainable agriculture practices. | Authorized the National Research Initiative Competitive Grants Program. | Congressional appropriators responded to the administration's initial request for $100 million by increasing the previous year's allocation of $42.5 million to $73 million. |
| 1994 | Elementary and Secondary Education Reauthorization Act | Conferred land grant status to the 29 Native American colleges that compose the American Indian Higher Education Consortium. | | For the 29 colleges a $23 million endowment, to be built up over a 5-year period, was authorized. The colleges would receive interest payments from the endowment each year. Also authorized were funds for the colleges' education and extension programs in agricultural and natural resources. |

SOURCES: Adapted from Alston, J. M., and P. G. Pardey. 1995. Making Science Pay: Economics of Financing, Organizing and Managing Public-Sector Agricultural R&D. Washington, D.C.: American Enterprise Institute. Augmented with information from Huffman, W. E., and R. E. Evenson. 1993. Science for Agriculture: A Long-Term Perspective. Ames: Iowa State University Press; and Christy, R. D., and L. Williamson, eds. 1992. A Century of Service: Land-Grant Colleges and Universities, 1890–1990. New Brunswick, N.J.: Transaction Publishers.

- Today, agricultural extension specialists are usually located at colleges of agriculture. They often have research appointments and, sometimes, teaching or teaching and research appointments. University-based extension specialists must interact with research scientists and relay scientific learning and other knowledge to farmers and other users. They also serve as the university's link to the county extension agents and the USDA's Extension Service.
- The tripartite mission—teaching, research, and extension—has been a hallmark of the land grant college of agriculture system. Over the years, however, divisive elements within the three-part mission have emerged. Teachers, researchers, and extension specialists often respond to different administrators, to different constituents with different interests, and to different incentives and awards.
- Over the decades a progression of legislative actions, as shown in Table 1-2, expanded funding to the college system, revamped funding mechanisms, expanded or refined the provisions for the use of federal funds, and even added institutions to the system. For example, the 1925 Purnell Act put a new emphasis on the system's role in improving rural home and rural life. The 1935 Bankhead-Jones Act established the original formula for allocating Hatch research funds among SAESs. The 1946 Research and Marketing Act revamped the formula and introduced a national advisory committee. The 1962 McIntire-Stennis Act created additional formula funds for forestry research. The 1977 National Agricultural Research, Extension, and Teaching Policy Act (the 1977 farm bill) instituted formula funds for research at 1890 colleges, formula funds for research programs in animal health, and a new competitive grants program to be administered by USDA but open to all scientists in and outside of the land grant system. The 1990 Food, Agriculture, Conservation, and Trade Act (the 1990 farm bill) expanded the competitive grants program of the 1977 act by mandating the National Initiative for Research on Agriculture, Food, and Environment (NRI). Most recently the 1994 Elementary and Secondary Education Act conferred land-grant status on 29 Native American colleges and authorized funding for their education and extension programs in agriculture and natural resources.

Schools and colleges of forestry and veterinary medicine, usually located at land grant universities, augment the college of agriculture system.

- Colleges of veterinary medicine began their affiliation with land grant universities in 1879 with the opening of the veterinary college at Iowa State University. Today, of the 27 veterinary colleges only 2—those at the University of Pennsylvania and at Tufts University—are not affiliated with land grant schools. Of the remaining 25, nine were established after 1967. (Table 1-1 lists and Figure 1-2 maps the veterinary medicine schools and colleges.)
- The majority of states that do not have colleges of veterinary medicine, and some that do, maintain significant programs in veterinary science in departments in colleges of agriculture.

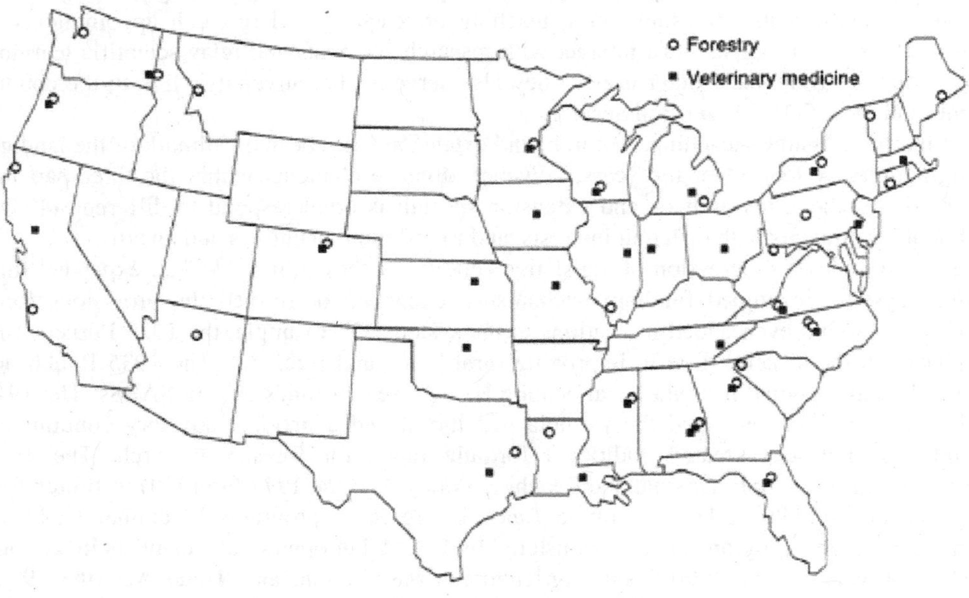


**FIGURE 1-2**
*Map shows locations of administratively separate schools and colleges of forestry and veterinary medicine. Other forestry and veterinary medicine programs are subunits of colleges of agriculture.*

- There are a number of links, actual and potential, between colleges of agriculture and veterinary medicine. Some faculty of veterinary medicine colleges have SAES appointments. These two types of colleges have overlapping interests in animal health research, and both have access to animal health research funds administered by USDA. Many veterinary medicine students receive their prior training in animal science departments at colleges of agriculture. Both often house and manage federal-state cooperative extension programs.
- Forestry programs are also linked to colleges of agriculture. They are located in independent forestry schools or colleges and in forestry departments in colleges of agriculture. There are more than 60 forestry programs in total. Most forestry programs are at land grant universities, though there are some prominent exceptions, such as those at the University of Washington, University of Michigan, the State University of New York at Syracuse, Yale University, Duke University, and at California's Humboldt State University. (Table 1-1 lists and Figure 1-2 maps the administratively separate forestry schools and colleges. Table 1-1 also indicates which land grant colleges of agriculture have forestry programs.)
- The passage of the McIntire-Stennis Act in 1962 (see Table 1-2), which made federal funds available for forestry research on a formula basis, spawned more than one-half of the current forestry programs. These funds are channeled to colleges of agriculture through SAESs and to forestry colleges and schools in and outside of the land grant system. However, the much larger amount of forestry research dollars is a component of the USDA Forest Service's budget.
- Although forestry and agricultural research and education are now often conducted in isolation, some argue the case for stronger program integration focused on ecosystem and landscape management—approaches that account for the interactions among farming, forestry, wildlife habitat, urbanization, and other land uses. One reason to pursue an integrated approach to research and education is that farmers own 82 million acres, or 17 percent, of all U.S. timberland (Powell et al., 1993).

## ISSUES FOR DISCUSSION

- The federal government has had a long and special role in the land grant college of agriculture system. Is there a continuing role for federal legislation in influencing the future missions and structure of the college of agriculture system, and what form should it take?
- Do the components of the current system—including colleges of agriculture, home economics, forestry, and veterinary medicine—operate together efficiently to deliver education, research, and technology development? For example, what institutional or curriculum changes might promote programs that more explicitly take account of interactions between commodity production and natural resource or forestry management?
- As land grant colleges have evolved into total universities, how have colleges of agriculture ensured that they are an integrated part of the larger university?
- The 1890 institutions have their own special legislative history and appropriations. Do 1890s have a unique role today? How are their functions and activities supported by and linked to those of the 1862 colleges?

## SUGGESTED READINGS

Christy, R. D., and L. Williamson, eds. *A Century of Service: Land-Grant Colleges and Their Universities, 1890-1990.* New Brunswick, N.J.: Transaction Publishers, 1992.

Kerr, N. A. *The Legacy: A Centennial History of the State Agricultural Experiment Stations: 1887-1987.* Columbia: University of Missouri, 1987.

Mayberry, B. D. *A Century of Agriculture in the 1890 Land-Grant Institutions and Tuskegee University—1890–1990.* New York: Vantage Press, 1991.

National Research Council. *Forestry Research: A Mandate for Change* , Washington, D.C.: National Academy Press, 1990.

Pritchard, William R., ed. *Future Directions for Veterinary Medicine* . Durham, N.C.: Duke University, Pew National Veterinary Education Program, 1989.

Rasmussen, Wayne D. *Taking the University to the People: Seventy-five Years of Cooperative Extension.* Ames:Iowa State University Press, 1989.

Schor, J. *Agriculture in the Black Land-Grant System to 1930.* Tallahassee: Florida A&M University, 1982.

# 2

# U.S. AGRICULTURE YESTERDAY AND TODAY: The Colleges' Changed Environment

*This chapter describes some of the main ways in which the U.S. farm and agricultural industry has evolved since the colleges' early years and some of the ways in which the colleges have contributed to those changes through science and technology development. The chapter's main goal is to provide an economic backdrop for a discussion of the current and evolving role of the colleges of agriculture in addressing society's needs and concerns. It draws heavily on data and reports generated by the U.S. Department of Agriculture's Economic Research Service (ERS).*

- In 1860 at the dawn of the decade that would put the land grant college system in the history books, one-half of the U.S. population lived on farms and more than one-half of the labor force worked on them. The numbers of farms continued to rise until the 1920s. In the decades that followed, however, U.S. citizens left farming in massive numbers for other ways of life and alternative types of employment. By 1990 the farm population was less than one-third of what it had been in 1860, and by 1992 there were slightly fewer farms than there had been in 1860 (Table 2-1).
- It is important to understand that these trends, in addition to having changed the profile of the national landscape, are also indicators of economic progress. The same number of farms and farmers can feed vastly larger numbers of people today than 100 years ago. The fact that so many more people could be fed with relatively little farm labor input meant that farm workers became available to other industries—industries that taught them different skills and paid them higher wages. Essentially, the release of labor from farming fueled the growth of the rest of the U.S. economy, although this change did not come without significant adjustment costs for communities and families.

Farm productivity has improved since the inception of and as the result of land grant colleges of agriculture.

TABLE 2-1 Total U.S. Population Statistics Compared to Farm Population Statistics, 1840–1992

| | Number (millions) | | | Percent of U.S. Total | |
|---|---|---|---|---|---|
| Year | U.S. Population | Farms | Farm Population | Farm Population | Farm Labor Force |
| 1840 | 17.1 | NA | 9.1 | 53 | 69 |
| 1850 | 23.2 | 1.4 | 11.7 | 50 | 64 |
| 1860 | 31.4 | 2.0 | 15.1 | 48 | 58 |
| 1870 | NA | 2.7 | NA | NA | 47 |
| 1880 | 50.2 | 4.0 | 23.0 | 46 | 49 |
| 1890 | 62.9 | 4.6 | 26.4 | 42 | 43 |
| 1900 | 76.0 | 5.7 | 29.4 | 39 | 38 |
| 1910 | 92.0 | 6.4 | 32.1 | 35 | 31 |
| 1920 | 105.7 | 6.5 | 31.6 | 30 | 27 |
| 1930 | 122.8 | 6.3 | 30.4 | 25 | 21 |
| 1940 | 131.8 | 6.1 | 30.8 | 23 | 18 |
| 1950 | 151.1 | 5.4 | 25.1 | 17 | 11 |
| 1959 | 177.8 | 3.7 | 16.6 | 9 | 8 |
| 1960 | 180.7 | NA | 15.6 | 9 | 8 |
| 1969 | 202.7 | 2.7 | 10.3 | 5 | 4 |
| 1970 | 205.0 | NA | 9.7 | 5 | 4 |
| 1980 | 227.7 | NA | 6.1 | 3 | 3 |
| 1982 | 232.2 | 2.2 | 5.6 | 2 | 3 |
| 1990 | 249.9 | NA | 4.6 | 2 | 3 |
| 1992 | 255.4 | 1.9 | NA | NA | 3 |

NOTE: Data are compiled from various sources. Population and labor statistics for 1840 through 1959 are from Chronological Landmarks in American Agriculture, USDA, Nov 1990; 1960 through 1992 are from the Economic Report of the President, 1995; number of farms from the Census of Agriculture, U.S. Bureau of the Census. NA, data not available.

- The colleges of agriculture generated many of the scientific and management advances that contributed to the growth of productivity in U.S. agriculture. Such advances include hybrid seeds, improved farm and production management techniques, improved genetic stock of food animals, and sophisticated financial management strategies. Total factor productivity—that is, the output generated by all farm inputs working together—increased almost 150 percent between 1948 and 1991 (Table 2-2).
- Many studies of public investments in agricultural science, such as those listed in Table 2-3, show large economic payoffs. They find high rates of return to U.S. agricultural research and development, even though the range of estimates is large and methodological problems make accurate measurement very difficult (Alston et al., 1994).
- A significant portion of the gain in productivity in agriculture is the result of the substitution of "modern" inputs—mechanical and chemical—for labor hours and land area. For example, farm labor input decreased 65 percent from 1948 to 1991, while chemical inputs increased 176 percent over these same years (Table 2-2). The application of farm chemicals, combined with other yield-enhancing technologies such as improved crop varieties, has made it possible to produce more food and fiber on virtually the same amount of land. Yield-enhancing technologies have also helped the United States become the world's leading exporter of farm and agricultural products. The agricultural trade balance has been favorable every year since 1960, and agricultural products compose about 10 percent of total U.S. merchandise exports (Executive Office of the President, 1995).

TABLE 2-2 Productivity and Input Use in U.S. Agriculture (1982 = 100)

| Year | Total Factor Productivity | Selected Indexes of Input Use | |
|---|---|---|---|
| | | Farm Labor | Chemical |
| 1948 | 52 | 251 | 34 |
| 1950 | 50 | 237 | 43 |
| 1955 | 54 | 211 | 45 |
| 1960 | 60 | 163 | 58 |
| 1965 | 71 | 141 | 73 |
| 1970 | 77 | 119 | 76 |
| 1975 | 84 | 114 | 91 |
| 1980 | 82 | 108 | 131 |
| 1982 | 100 | 100 | 100 |
| 1985 | 110 | 89 | 101 |
| 1990 | 126 | 87 | 90 |
| 1991 | 124 | 88 | 94 |

NOTE: In computing productivity and input use indexes, 1982 is used as the base year. Farm output measures the annual volume of net farm production available for eventual human use through farm sales or on-farm consumption. Total factor productivity is measured by farm output per unit of total factor input.

SOURCE: Adapted from Executive Office of the President. 1994. Economic Report of the President. Washington, D.C.: U.S. Government Printing Office

- Despite public benefits, concerns about modern farming technologies, particularly the impacts of farm chemicals on human health, soil and water quality, and wildlife, have intensified in recent years. Mechanical technologies can also have adverse effects on soil quality and soil erosion. There is evidence that agricultural chemicals and sediments are impairing the quality of some surface and groundwater resources and imposing costs on water users (U.S. Department of Agriculture, Economic Research Service, Natural Resources and Environment Division, 1994). Thus agricultural research has been directed, more recently, toward developing production technologies that are both cost-effective substitutes for machines and chemicals and less risky in terms of environmental and health costs.
- Technologies developed and in use that can reduce reliance on mechanical and chemical inputs and, at the same time, enhance farm productivity include integrated pest management and other "best-management" practices, such as crop rotations with legumes, integrated livestock-crop systems with manure applications, and management-intensive grazing (see Vandeman et al., 1994, for an assessment of integrated pest management [IPM] adoption). Also, reliance on conventional chemical pest control is being reduced through the ongoing development of biological controls and through classical plant breeding methods that continue to improve crop resistance to insects and disease.
- Frontier developments in biotechnology may also offer the opportunity to achieve greater compatibility between farm productivity and environmental quality. Scientists in both the public and private sectors are working to apply bioengineering techniques to the development of pest-and disease-resistant crops. For example, advances in biological nitrogen fixation technologies could make plants more efficient in absorbing nitrogen and thereby reduce the need for synthetic nitrogen fertilizers derived from fossil fuels. Bioengineering techniques are also being directed toward improving crop tolerance to chemical herbicides, which will reduce crop loss from weeds but, some argue, could reinforce the use of chemical controls (Caswell et al., 1994).

**TABLE 2-3** Studies Reporting Estimated Annual Rates of Return on Investments in U.S. Agricultural Research and Development, 1958–1990

| Author | Year | Commodity | Period | Rate of Return (percent) |
|---|---|---|---|---|
| Griliches | 1958 | Hybrid corn | 1940–1955 | 35–40 |
| | | Hybrid sorghum | 1940–1957 | 20 |
| Griliches | 1964 | Aggregate | 1949–1959 | 35–40 |
| Latimer | 1964 | Aggregate | 1949–1959 | NS |
| Peterson | 1967 | Poultry | 1915–1960 | 21–25 |
| Evenson | 1986 | Aggregate | 1949–1959 | 47 |
| Schmitz and Seckler | 1970 | Tomato harvester | 1958–1969 | 37–46 |
| Huffman[a] | 1974 | Maize | 1959–1964 | >16 |
| Cline | 1975 | Aggregate | 1939–1948 | 41–50 |
| | | | 1949–1958 | 39–47 |
| | | | 1959–1968 | 32–39 |
| | | | 1969–1972 | 28–35 |
| Peterson and Bredahl | 1975 | Aggregate | 1937–1942 | 50 |
| | | | 1947–1957 | 51 |
| | | | 1957–1962 | 49 |
| | | | 1967–1972 | 34 |
| Bredahl and Peterson | 1976 | Cash grains | 1969 | 36 |
| | | Poultry | 1969 | 37 |
| | | Dairy | 1969 | 43 |
| | | Livestock | 1969 | 47 |
| Huffman[a] | 1977 | Crops | 1959–1964 | 110 |
| Peterson and Fitzharris | 1977 | Aggregate | 1937–1972 | 34-51 |
| | | | 1937–1942 | 50 |
| | | | 1947–1952 | 51 |
| | | | 1957–1962 | 49 |
| | | | 1957–1972 | 34 |
| Evenson | 1978 | Aggregate | 1948–1971 | 110 |
| Evenson et al. | 1979 | Aggregate | 1868–1926 | 65 |
| | | | 1927–1950 | 95–110 |
| | | | 1948–1971 | 45 |
| Knutson and Tweeten | 1979 | Aggregate | 1949–1972 | 28–47 |
| Lu, Cline and Quance | 1979 | Aggregate | 1939–1972 | 23.5–30.5 |
| White et al. | 1979 | Aggregate | 1929–1977 | 28–37 |
| Davis and Peterson | 1981 | Aggregate | 1949–1974 | 37–100 |
| Norton | 1981 | Cash grains | 1969 | 31–57 |
| | | Poultry | | 30–55 |
| | | Dairy | | 27–50 |
| | | Livestock | | 56–111 |
| Otto and Havlicek | 1981 | Corn | 1967–1979 | 152–210 |
| | | Wheat | | 79–148 |
| White and Havlicek | 1982 | Aggregate | 1943–1977 | 7–36 |
| Lyu, White and Liu | 1984 | Aggregate | 1949–1981 | 66–83 |
| Braha and Tweeten | 1986 | Aggregate | 1959–1982 | 47 |
| Huffman and Evenson | 1989 | All[b] | 1950–1982 | 43 |
| | | Crops[b] | 1950–1982 | 45 |
| | | Livestock[b] | 1950–1982 | 11 |
| | | All[c] | 1950–1982 | 67 |
| | | Crops[c] | 1950–1982 | 57 |
| | | Livestock[c] | 1950–1982 | 83 |
| | | All[d] | 1950–1982 | 83 |
| | | Crops[d] | 1950–1982 | 90 |
| | | Livestock[d] | 1950–1982 | 71 |

NOTE: NS, not statistically significant.

[a]Huffman (1974) and Huffman (1977) cover the corn-belt regions. The regional coverage of all other studies is national.
[b]Public-sector applied research.
[c]Public-sector pre-technology science.
[d]Private-sector research.

SOURCE: Adapted from Alston, J. M., P. G. Pardey, and H. O. Carter, eds. 1994. Valuing UC Agricultural Research and Extension. Agricultural Issues Center Pub. No. VR-1. Davis: University of California, Division of Agricultural and Natural Resources.

TABLE 2-4 Dollars (billions) Spent on Food Consumption and U.S. Farm Value's Share

| | Personal Consumption Expenditures | | | |
|---|---|---|---|---|
| Year | Total | All Food | Domestically Produced Food | Farm-Level Value of Food Produced and Consumed in the U.S. |
| 1950 | 192.1 | 53.9 (28.1) | 44.0 | 18.0 (41.0) |
| 1960 | 332.4 | 82.6 (24.8) | 66.9 | 22.3 (33.3) |
| 1970 | 646.5 | 142.1 (22.0) | 110.6 | 35.5 (32.1) |
| 1980 | 1,748.1 | 341.8 (19.6) | 264.4 | 81.7 (30.9) |
| 1990 | 3,761.2 | 604.8 (16.1) | 449.8 | 106.2 (23.6) |
| 1994 | 4,627.0 | 679.1[a] (15.0) | 510.6 | 109.6 (21.5) |

NOTE: Numbers in parentheses under "All Food" show percent of total personal consumption expenditures. Numbers in parentheses under "Farm Level Value ..." show percent of personal consumption expenditures for domestically produced food.

[a] Preliminary.

SOURCE: Total and all food personal consumption expenditures are from the Economic Report of the President, 1995 (Executive Office of the President. 1995. Budget of the United States Government. Washington, D.C.: U.S. Government Printing Office). Personal consumption expenditures for domestically produced food and farm value are from USDA, ERS, Food Cost Review, AER No. 696, 1993 and personal communication form 1994 update.

Farm-productivity gains translate into lower food costs.

- Research targeted to improve farm productivity has benefited consumers in both the United States and other countries. Increased productivity means that the same farm output can be produced at lower production cost; thus the cost of farm commodities to industries that manufacture food and fiber products (or to countries that import farm commodities) is reduced. These savings are passed on, at least partially, to final consumers.
- The consumer benefit translates into U.S. families spending, on average, a relatively small share of their personal consumption expenditures on food. That average share is now 15 percent, down from 28 percent in 1950 (Table 2-4). In other words, over the decades total personal consumption of all goods and services has increased much faster than has personal expenditures on food. As concern with food costs has lessened, other consumer concerns have gained visibility. Food safety, the amount of fats and cholesterol in food products, and the application of biotechnology to food production are currently food issues that concern U.S. consumers (see box copy, p. 23).
- Affordability of food is still an issue for low-income consumers. Insufficient family income and inadequate food distribution and access are often more at issue, however, than high farm-commodity costs. Because of inadequate access from inner city and poor rural areas to competitively priced retail outlets, food prices in low-income areas tend to be 20 to 36 percent higher than in higher-income areas (McGrath Morris et al., 1992; Troutt, 1993). It is also the case that all U.S. consumers pay higher-than-free-market prices for sugar, dairy, and peanut products because of government programs designed to support prices to domestic producers.

**What Americans Think about Food Issues**

A range of polls support the finding that consumers care about food issues. Stated consumer concerns may be in line with or different than scientifically supported knowledge about health and safety risks. There is incomplete evidence regarding the impact of consumer concerns on food purchases and consumption.

According to a nationwide survey commissioned by Public Voice for Food and Health Policy, a national nonprofit advocacy organization, concern among U.S. citizens about the effects agricultural chemicals have on health and the environment is very strong and widespread. The poll, conducted in 1993 by Fingerhut/ Granados Opinion Research Company, found that a majority of U.S. citizens were "very concerned" about how chemicals used to grow food affect the health of young children (68 percent); health problems caused generally by chemicals and pesticides used to grow food (60 percent); pesticides and fertilizers getting into the water supply (71 percent); and the risk of severe food poisoning from bacteria in meat (61 percent). The Public Voice poll is interesting in that it contrasts food-related health and environmental concerns with other health and environmental concerns. The poll found somewhat smaller percentages of people "very concerned" about health problems caused by secondary smoke—that is, smoke from other peoples' cigarettes (55 percent); health problems caused by air pollution from cars and industries (47 percent); and the effects of antibiotics and growth hormones used in meat and milk products (54 percent).

The Center for Produce Quality (CPQ), a nonprofit foundation created by the Produce Marketing Association and the United Fresh Fruit and Vegetable Association, found in a 1992 nationwide poll that adults were generally confident about the safety of fresh fruits and vegetables, but that 61 percent were nonetheless "very concerned" about pesticide residues. CPQ found that between 1989 and 1992 increased concern about pesticide residues was paralleled by growing consumer concerns about virtually all food-related issues, including nutritional value, fat, salmonella, cholesterol, and animal growth hormones.

The International Food Information Council of the American Dietetic Association found in 1993 that 44 percent of adults surveyed were "very concerned" about the effects their diet has on their health and that an additional 40 percent were "fairly concerned." In addition, they found that a strong majority of adults agreed that there are too many conflicting reports about nutrition.

A 1993 poll of 1,000 shoppers, conducted by the Food Marketing Institute, found that taste was the most important consideration when selecting food. Ninety-one percent of the shoppers considered taste "very important." Other factors were also ranked "very important" by a majority of shoppers including nutrition (75 percent), price (74 percent), and product safety (72 percent). The same study found that food attributes considered by a majority of the shoppers polled to be "serious hazards" included residues such as pesticides and herbicides (79 percent) and antibiotics and hormones in poultry and livestock (55 percent). Other attributes considered "something of a hazard" included nitrites in food, irradiated foods, additives and preservatives, and artificial coloring.

SOURCES: The following unpublished reports were used: Morris, P. M., A. Rosenfeld, and M. Bellinger. 1993. What Americans Think About Agrichemicals: A Nationwide Survey on Health, the Environment, and Public Policy. Washington, D.C.: Public Voice for Food and Health Policy; Center for Produce Quality. Produce Confidence, Consumption Grow. Alexandria, VA: Center for Produce Quality; American Dietetic Association. 1994. How Are Americans Making Food Choices? Washington, D.C.: International Food Information Council, American Dietetic Association; Food Marketing Institute. 1993 Trends. Washington, D.C.: Food Marketing Institute.

- Many U.S. citizens may have little sense of the continuing benefits they receive as a result of farm productivity-enhancing research. One reason the benefits are not readily apparent is that the raw-product component of retail food costs is so small. The effects of lower wheat prices on food prices in the supermarket, for example, may be all but unobservable after processing, packaging, marketing, shipping, and retail costs are added on; and these beyond-the-farm-gate costs have risen substantially over the years. U.S. farmers received only 21 percent of what U.S. consumers spent for domestically produced food in 1994, compared with 41 percent in 1950 (Table 2-4). The increasing share of food consumer away from home further increases the gap between farm-level commodity prices and retail food costs.

## THE CHANGING STRUCTURE OF U.S. AGRICULTURE

- Over the decades, although the number of U.S. farms decreased, the amount of land used for farming stayed more or less the same. As individual farms got larger they also became considerably more unequal in their contributions to national farm output. Concentration of commercial production is perhaps the most striking feature of modern U.S. agriculture. Today, only 3.6 percent of all farms account for one-half the value of all farm output, and 1.5 percent of all farms account for one-third of all output (Table 2-5). Of the approximately 2 million farms, about 280,000 provide most of the food and fiber that enter commercial channels. Although the contribution of agricultural research to farm-sector concentration is uncertain, what is clear is that relatively little research has been directed toward understanding the causes of this trend or its social implications or effects on the food system.
- In addition to the fact that only a small fraction of all farms account for most farm output is the fact that most farm households do not rely on farm sales for most of their household income. A recent ERS report (Hoppe, 1994) examined farm businesses and farm operator households in county groupings labeled as follows:
- farming-dependent counties (20 percent of local earnings come from farming),
- major farming counties (less than 20 percent of local earnings come from farming, but farms in these counties rank in the top 20 percent of U.S. counties in total farm earnings), and
- residual counties (all other U.S. counties, including metropolitan counties).

TABLE 2-5 Total U.S. Farms and Concentration of Farm Output, 1990-1992

| | | Farms Accounting for One-Half of Output[a] | | Farms Accounting for One-Third of Output | |
|---|---|---|---|---|---|
| Year | Number of Farms | Percent of All Farms | Average Acres per Farm | Percent of All Farms | Average Acres per Farm |
| 1900 | 5,751,830 | 17.1 | 369 | NA | NA |
| 1940 | 5,938,897 | 11.6 | 611 | 5.2 | 989 |
| 1969 | 2,736,914 | 8.1 | 1,611 | 1.9 | 3,305 |
| 1987 | 2,102,278 | 3.6 | 2,792 | 1.5 | 3,921 |
| 1992 | 1,925,300 | NA | NA | NA | NA |

NOTE: Output is measured as sales. NA, data not available.

[a] Includes farms accounting for one-third of all output.

SOURCE: Peterson, R., and N. Brooks. 1993. The Changing Concentration of U.S. Agricultural Production During the 20th Century: 14th Annual Report to the Congress on the Status of the Family Farm. Agriculture Information Bulletin No. 671. Washington, D.C.: U.S. Department of Agriculture, Economic Research Service.

TABLE 2-6 Financial Characteristics of Farm-Operator Households, by County Group, 1990

| | County Groups | | | |
|---|---|---|---|---|
| Variable | Farming-Dependent | Major Farming | Residual | Total |
| Number of farm-operator households | 229,811 | 424,762 | 1,083,446 | 1,738,019 |
| Household income ($ per household) | $40,413 | $52,624 | $33,370 | $39,007 |
| Farm-related income | 15,127 | 10,042 | 2,066 | 5,742 |
| Off-farm income[a] | 25,286 | 42,582 | 31,304 | 33,265 |
| Wages and salaries | 12,942 | 19,298 | 17,239 | 17,174 |
| Interest or dividends | 2,483 | 4,494 | 2,846 | 3,201 |
| Other off-farm income[b] | 4,269 | 6,226 | 5,133 | 5,286 |
| Negative income (% of households) | | | | |
| Farm-related income | 38.4 | 53.9 | 59.5 | 55.3 |
| Total household income | 9.6 | 11.2 | 7.4 | 8.6 |
| Farm income compared with off-farm income (% of households) | | | | |
| No off-farm income | 11.0 | 10.0 | 6.8 | 8.1 |
| Farm income less | 60.5 | 71.3 | 81.8 | 76.4 |
| Farm income equal or greater | 28.5 | 18.7 | 11.4 | 15.5 |
| Net worth of farm operated ($ per household)[c] | $342,215 | $461,407 | $278,308 | $331,506 |

NOTE: A "farm-operator household" is one that either works on the farm of makes day-to-day decisions about such things as planting, harvesting, feeding, and marketing. It may share the net worth of the farm with one or more other nonoperator farm households.

[a] Includes off-farm business income not shown separately.

[b] Net income from estates and trusts, rental income from nonfarm properties, royalties from mineral leases, retirement/disability income, annuities, alimony, regular contributions from persons not in the household, and any other miscellaneous sources of income.

[c] Net worth may be shared with nonoperator farm households.

SOURCE: Hoppe, R. A. 1994. Farming Operations and Households in Farming Areas: A Closer Look. Agricultural Economic Report No. 685. Washington, D.C.: U.S. Department of Agriculture, Economic Research Service.

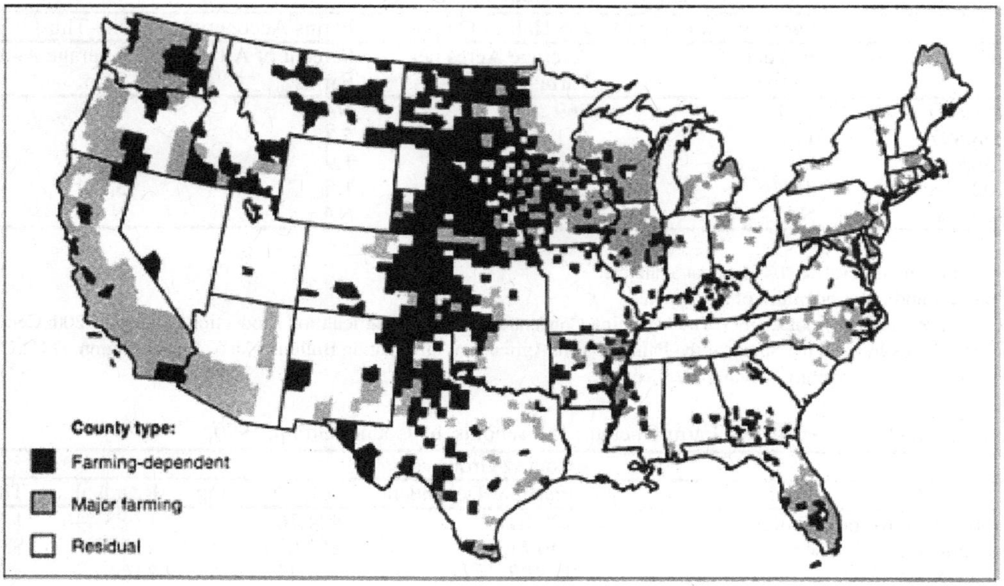


**FIGURE 2-1**

*Dependence on income from farming is shown according to county types: farming-dependent, major farming, and residual. SOURCE: Hoppe, R. A. 1994. Farming Operations and Households in Farming Areas: A Closer Look. Agricultural Economic Report No. 685. Washington, D.C.: Economic Research Service, U.S. Department of Agriculture.*

- Table 2-6 shows that for almost 80 percent of farm-operator households, farm-related income is less than off-farm income. Off-farm income is least important (in relation to farm-related income) in farming-dependent counties, which are located predominately in the northern Midwest and Great Plains states and some parts of the Northwest (Figure 2-1).
- Specialized production is another characteristic of the modern farm economy. Whereas decades ago farms may have been integrated production units producing a variety of crop and livestock products to meet home and local market needs, today they usually specialize in products that represent their region's "comparative advantage" in national and international markets. For example, more than 70 percent of total U.S. sales of corn, soybeans, and hogs derive from several midwest "corn belt" states; about 70 percent of poultry is produced in counties concentrated in six or seven southeastern states; and about 70 percent of tobacco is produced in counties concentrated in four to five eastern states (Table 2-7 and Figure 2-2).
- In addition, the farm economies of some regions depend heavily on their regional production specialties. For example, the Great Plains counties that specialize in cattle, wheat, and sorghum count on these commodities for more than 80 percent of their farm sales; the poultry-specializing counties count on poultry for almost 70 percent of farm sales; and the dairy-specializing counties of the Northeast and Great Lakes states depend on dairy product sales for more than 55 percent of their farm sales (Table 2-7 and Figure 2-2).

TABLE 2-7 Production Specialization in the U.S. Farm Economy

| Production Specialty/ Specialty Group | Number of Specializing Counties | Percent of Total County-Cluster Farm Sales | Percent of Total U.S. Sales of Commodity/Commodities |
|---|---|---|---|
| Corn/soybeans/hogs | 508 | 64 | 72/73/71 |
| Poultry | 248 | 66 | 68 |
| Dairy | 164 | 56 | 42 |
| Gattle/wheat/sorghum | 343 | 81 | 37/41/65 |
| Tobacco | 135 | 37 | 69 |
| Cattle (part-time) | 371 | 56 | 11 |
| Fruit | 26 | 47 | 60 |
| Other crops | 116 | 27 | 47 |
| Vegetables/nursery | 86 | 38 | 63/70 |
| Wheat/oats/other grain | 115 | 51 | 22/19/42 |
| Cotton | 103 | 45 | 59 |
| Sheep/cattle/other livestock | 75 | 73 | 39/4/32 |
| Total counties | 2,290 | | |

Across the nation, agricultural interests and concerns vary from one constituent group to another and from one state to another.

- Vertical integration—a system that combines previously separated stages of the production and delivery system in a single firm—is increasing in some areas of the United States, particularly areas specializing in swine and poultry production. Vertical integration can take a number of forms but may typically involve contractual relationships between the "integrators" and the producers. The trend toward vertical integration results from a number of factors including advances in food technology and greater globalization of agricultural production and trade. In the case of pork, for example, the process of engineering the final product to meet specific consumer demands begins at the hog production level. Vertical integration also assures processors a steady supply of inputs so that they can consistently utilize their plants at optimum capacity and seek expansions in product markets (Council on Food, Agricultural, and Resource Economics, 1994).
- The above described changes in U.S. farming do not diminish the economic importance of the agriculture complex to the United States. The food and fiber sector of the domestic economy is large, accounting for 18 percent of U.S. employment and about 16 percent of "value added" to domestic production (Table 2-8). Most employment and value added production occurs beyond the farm gate—in food processing, manufacturing, transportation, and retailing in stores and in restaurants—but farm production underpins all these non-farm activities. Furthermore, the forestry sector, which is intertwined with farming in some parts of the country, accounts for another 5 percent of the value of national economic output (U.S. Department of Agriculture, 1990a).

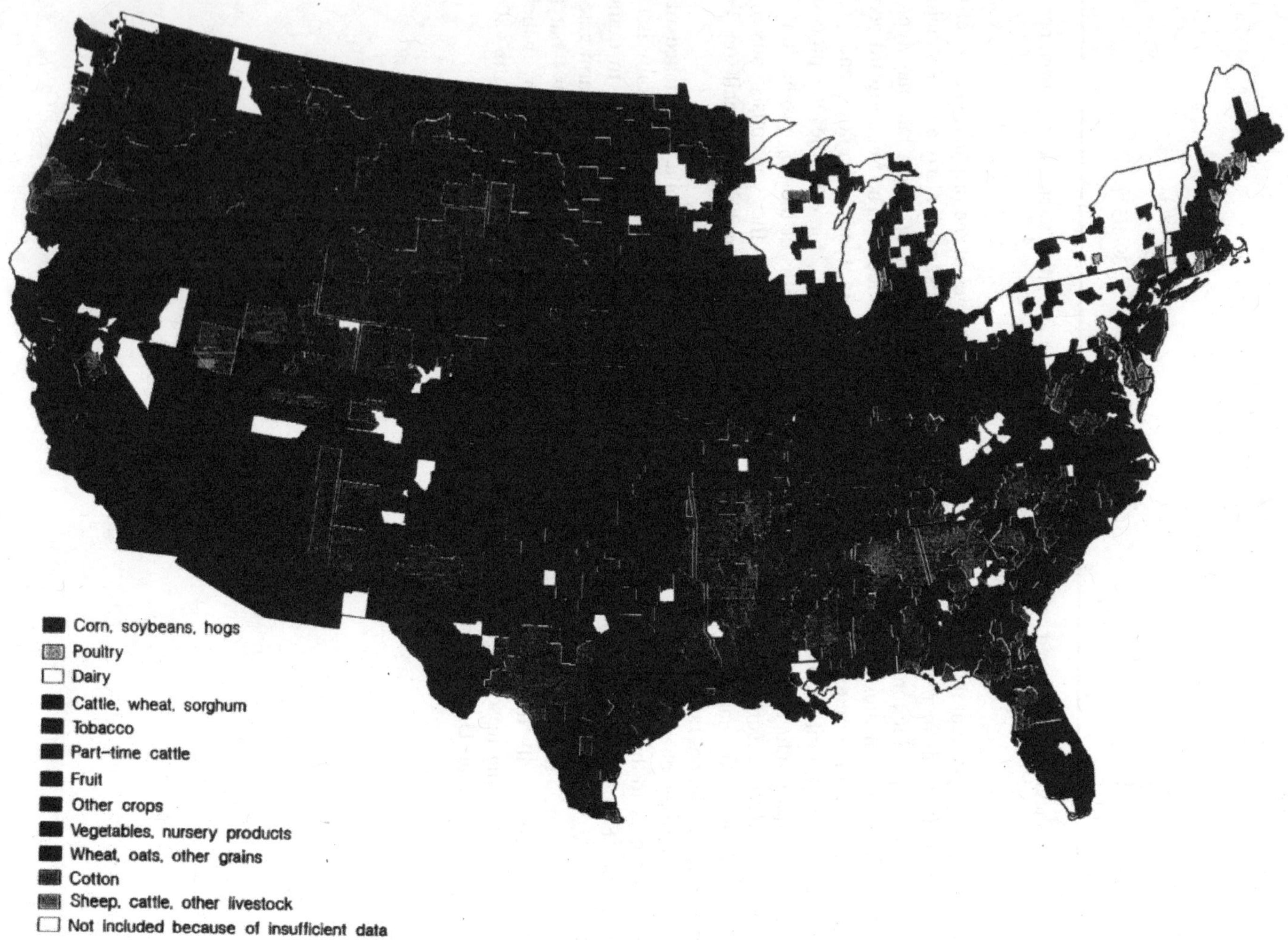


**FIGURE 2-2**

*Map shows patterns of agricultural specialization, by multicounty cluster, based on twelve individual commodities or commodity groups. SOURCE: Sommer, J. E., and F. K. Hines. 1991. Diversity in U.S. Agriculture: A New Delineation by Farming Characteristics, AER No. 646. Washington, D.C.: Economic Research Service, U.S. Department of Agriculture.*

**TABLE 2-8** The Food and Fiber Sector's Contribution to the U.S. Economy, 1982–1992

| | Year | | | | | | | | | | |
|---|---|---|---|---|---|---|---|---|---|---|---|
| Item | 1982 | 1983 | 1984 | 1985 | 1986 | 1987 | 1988 | 1989 | 1990 | 1991 | 1992 |
| | *Millions of Workers* | | | | | | | | | | |
| Employment | | | | | | | | | | | |
| Total food and fiber | 22.9 | 23.0 | 22.6 | 22.3 | 22.4 | 22.4 | 22.7 | 23.4 | 23.5 | 23.2 | 22.8 |
| Farm sector | 2.4 | 2.8 | 2.3 | 2.0 | 2.0 | 2.0 | 2.0 | 2.0 | 2.1 | 2.0 | 2.0 |
| Nonfarm sectors | 20.5 | 20.2 | 20.3 | 20.2 | 20.3 | 20.5 | 20.6 | 21.4 | 21.4 | 21.2 | 20.8 |
| Food processing | 1.7 | 1.6 | 1.6 | 1.6 | 1.6 | 1.6 | 1.5 | 1.5 | 1.5 | 1.5 | 1.5 |
| Manufacturing | 3.6 | 3.5 | 3.4 | 3.3 | 3.2 | 3.1 | 3.0 | 3.2 | 3.1 | 3.1 | 3.0 |
| Transportation, trade and retailing | 7.2 | 7.2 | 7.3 | 7.4 | 7.5 | 7.5 | 7.6 | 7.9 | 8.0 | 7.9 | 7.8 |
| Eating | 4.6 | 4.7 | 4.7 | 4.7 | 4.8 | 5.0 | 5.2 | 5.3 | 5.4 | 5.3 | 5.2 |
| All other | 3.4 | 3.2 | 3.2 | 3.2 | 3.2 | 3.2 | 3.3 | 3.5 | 3.5 | 3.5 | 3.4 |
| Total domestic economy | 110.2 | 111.6 | 113.5 | 115.5 | 117.8 | 119.9 | 121.7 | 123.9 | 124.8 | 125.3 | 127.0 |
| | *Percent of Total U.S. Labor Force* | | | | | | | | | | |
| Farm sector | 2.1 | 2.5 | 2.0 | 1.8 | 1.7 | 1.6 | 1.7 | 1.6 | 1.7 | 1.6 | 1.6 |
| Nonfarm sectors | 17.5 | 16.7 | 16.5 | 17.5 | 17.3 | 17.1 | 17.0 | 17.3 | 17.2 | 16.9 | 16.4 |
| Total domestic economy | 100.0 | 100.0 | 100.0 | 100.0 | 100.0 | 100.0 | 100.0 | 100.0 | 100.0 | 100.0 | 100.0 |
| | *$ Billions* | | | | | | | | | | |
| Value added by activity | | | | | | | | | | | |
| Total food and fiber | 622.6 | 630.5 | 672.1 | 697.8 | 716.6 | 744.7 | 784.5 | 850.3 | 889.8 | 913.8 | 950.2 |
| Farm sector | 58.7 | 40.9 | 55.5 | 58.5 | 52.6 | 53.0 | 49.7 | 65.2 | 68.1 | 63.4 | 67.0 |
| Nonfarm sectors | 563.9 | 589.6 | 616.6 | 639.3 | 664.0 | 691.7 | 734.9 | 785.0 | 821.7 | 850.4 | 883.2 |
| Food processing | 72.6 | 76.9 | 80.5 | 82.3 | 87.4 | 86.2 | 91.8 | 96.1 | 99.8 | 103.4 | 106.2 |
| Manufacturing | 105.0 | 108.3 | 109.9 | 109.8 | 113.2 | 111.5 | 115.2 | 124.5 | 125.6 | 130.8 | 135.7 |
| Transportation, trade and retailing | 187.2 | 193.2 | 201.2 | 208.9 | 212.0 | 222.5 | 235.7 | 247.2 | 257.4 | 265.6 | 277.5 |
| Eating | 57.4 | 60.5 | 62.9 | 64.5 | 66.0 | 73.0 | 78.5 | 80.7 | 85.6 | 88.3 | 90.9 |
| All other | 141.8 | 150.7 | 162.1 | 173.8 | 185.3 | 198.4 | 213.8 | 236.6 | 253.2 | 262.3 | 272.9 |
| Total domestic economy | 3,149.6 | 3,405.0 | 3,777.2 | 4,038.7 | 4,268.6 | 4,539.9 | 4,900.4 | 5,250.8 | 5,546.1 | 5,722.9 | 6,038.5 |
| | *Percent of Total U.S. Economy* | | | | | | | | | | |
| Farm sector | 1.9 | 1.2 | 1.5 | 1.4 | 1.2 | 1.2 | 1.0 | 1.2 | 1.2 | 1.1 | 1.1 |
| Nonfarm sectors | 17.9 | 17.3 | 16.3 | 15.8 | 15.6 | 15.2 | 15.0 | 15.0 | 14.8 | 14.9 | 14.6 |
| Total domestic economy | 100.0 | 100.0 | 100.0 | 100.0 | 100.0 | 100.0 | 100.0 | 100.0 | 100.0 | 100.0 | 100.0 |

SOURCE: Data are from USDA Economic Research Service (ERS).

- The importance of agriculture to other aspects of U.S. life, though harder to measure, is no less important. The relationship between diet and human health, for example, is increasingly recognized by scientists and the general public. Heart disease, cancer, stroke, and diabetes—the four leading causes of death in the United States—have been linked to diet. According to some research, proper diet might forestall at least 20 percent of deaths, annually, from these four causes; other factors include genetic predisposition, smoking, and exercise. Hypertension, osteoporosis, and obesity, which affect productivity and life span, are also diet-related (Frazão, 1995). Also, increasingly recognized is the fact that farmers own or manage the majority of privately held land and thus de facto manage the associated natural resources including watersheds and wildlife.
- The ways in which states differ may be significant to the future of the colleges of agriculture. Characteristics of states' farm sectors differ; the role of agriculture in state economies differs, as do the interfaces between agriculture and other state land uses. Characteristics of states' population sectors differ; the importance of rural, suburban, and urban constituencies varies as do the income levels and age and ethnic demographics of each state's communities.

**TABLE 2-9** Total Cash Receipts (thousands of dollars) from Farming, 1990

| State | Cash Receipts from Farm Marketings | Government Payments | Cash Receipts from Farming |
|---|---|---|---|
| California | 19,205,854 (3) | 252,333 (<1) | 19,458,187 (3) |
| Texas | 11,826,661 (3) | 974,702 (<1) | 12,801,363 (3) |
| Iowa | 10,273,202 (18) | 753,733 (1) | 11,026,935 (20) |
| Nebraska | 8,715,014 (26) | 624,646 (2) | 9,339,660 (28) |
| Illinois | 7,767,202 (3) | 506,603 (<1) | 8,273,805 (3) |
| Kansas | 7,019,018 (14) | 834,746 (2) | 7,853,764 (15) |
| Minnesota | 6,888,461 (7) | 511,759 (1) | 7,400,220 (7) |
| Wisconsin | 5,718,984 (6) | 181,243 (<1) | 5,900,227 (6) |
| Florida | 5,717,193 (2) | 37,155 (<1) | 5,754,348 (2) |
| Indiana | 4,907,626 (4) | 244,170 (<1) | 5,151,796 (5) |
| North Carolina | 4,962,498 (4) | 73,255 (<1) | 5,035,753 (4) |
| Ohio | 4,425,163 (2) | 197,006 (<1) | 4,622,169 (2) |
| Arkansas | 4,251,574 (11) | 312,696 (1) | 4,564,270 (12) |
| Colorado | 4,218,122 (6) | 236,723 (<1) | 4,454,845 (6) |
| Missouri | 3,984,520 (4) | 299,065 (<1) | 4,283,585 (4) |
| Georgia | 3,857,804 (3) | 130,593 (<1) | 3,988,397 (3) |
| Washington | 3,752,119 (3) | 205,425 (<1) | 3,957,544 (4) |
| Oklahoma | 3,548,467 (6) | 319,040 (1) | 3,867,507 (7) |
| Pennsylvania | 3,680,180 (2) | 41,414 (<1) | 3,721,594 (2) |
| South Dakota | 3,260,859 (25) | 332,851 (3) | 3,593,710 (28) |
| Michigan | 3,111,876 (2) | 168,831 (<1) | 3,280,707 (2) |
| Kentucky | 3,102,981 (1) | 81,610 (<1) | 3,184,591 (1) |
| North Dakota | 2,531,265 (21) | 545,378 (5) | 3,076,643 (26) |
| New York | 2,966,160 (1) | 59,304 (<1) | 3,025,464 (1) |
| Idaho | 2,800,678 (15) | 133,431 (1) | 2,934,109 (15) |
| Alabama | 2,827,260 (4) | 82,226 (<1) | 2,909,486 (4) |
| Mississippi | 2,432,587 (6) | 185,969 (<1) | 2,618,556 (7) |
| Oregon | 2,371,412 (4) | 89,137 (<1) | 2,460,549 (4) |
| Virginia | 2,151,334 (2) | 32,378 (<1) | 2,183,712 (2) |
| Tennessee | 2,056,909 (2) | 91,029 (<1) | 2,147,938 (2) |
| Louisiana | 1,915,938 (2) | 154,631 (<1) | 2,070,569 (2) |
| Montana | 1,653,394 (13) | 299,599 (2) | 1,952,993 (15) |
| Arizona | 1,908,577 (3) | 43,349 (<1) | 1,951,926 (3) |
| New Mexico | 1,483,465 (5) | 63,840 (<1) | 1,547,305 (6) |
| Maryland | 1,359,502 (1) | 17,386 (<1) | 1,376,888 (1) |
| South Carolina | 1,168,677 (2) | 62,637 (<1) | 1,231,314 (2) |
| Wyoming | 756,694 (6) | 31,283 (<1) | 787,977 (6) |
| Utah | 745,154 (2) | 34,897 (<1) | 780,051 (3) |
| New Jersey | 648,791 (<1) | 15,744 (<1) | 664,535 (<1) |
| Delaware | 635,845 (3) | 3,213 (<1) | 639,058 (3) |
| Hawaii | 600,049 (2) | 519 (<1) | 600,568 (2) |
| Maine | 491,917 (2) | 6,982 (<1) | 498,899 (2) |
| Connecticut | 473,890 (1) | 2,123 (<1) | 476,013 (1) |
| Vermont | 457,167 (4) | 5,793 (<1) | 462,960 (4) |
| Massachusetts | 443,395 (<1) | 3,023 (<1) | 446,418 (<1) |
| West Virginia | 332,997 (1) | 6,049 (<1) | 339,046 (1) |
| Nevada | 324,532 (1) | 5,347 (<1) | 329,879 (1) |
| New Hampshire | 142,282 (1) | 1,856 (<1) | 144,138 (1) |
| Rhode Island | 71,346 (<1) | 191 (<1) | 71,537 (<1) |
| Alaska | 26,663 (<1) | 1,117 (<1) | 27,780 (<1) |
| Total | 169,973,258 (3) | 9,298,030 (<1) | 179,271,288 (3) |

NOTE: Cash receipts from farm marketings plus government subsidy payments equal total cash receipts from farming. Numbers in parentheses are percent of gross state product.

SOURCE: U.S. Department of Agriculture. 1990. State Financial Summary. Washington, D.C.: U.S. Department of Agriculture.

- Each state's unique blend of characteristics causes each to contribute differently to the U.S. farm economy. Three states, California, Texas, and Iowa, account for nearly one-quarter of the national value (cash receipts) of farm marketings (Table 2-9). These together with Nebraska and Illinois yield more than one-third of total marketed U.S. farm output, although other states, as noted above, are vitally important for specific types of crop or livestock production.
- Despite the fact that these states are major contributors to national farm output, for California, Texas, and Illinois farm marketing receipts amount to only 3 percent of gross state product. On the other hand, in Iowa and Nebraska farm marketings equal 18 and 26 percent, respectively, of the value of gross state product. Across the country, only eight states generate more than 10 percent of their gross state product from farming, with Nebraska, North Dakota, and South Dakota (wheat producing states) being the most dependent on farm income (Table 2-9). (For a discussion of methodological issues in developing conceptually consistent measures of the role of agriculture in state economies, see Leones et al., 1994.)
- Across the nation farming rarely provides significant state employment opportunities in relation to other industries. Only North and South Dakota employ more than 10 percent of their labor force in farm jobs. However, many if not most states look to farm-related industries, mostly the wholesale and retail trades, to provide a significant share of state employment. The agricultural processing and marketing industry is particularly important to employment in several southern states (Table 2-10).
- Today the population of most states is significantly urban, and the farm sector has decreased proportionately; however, the size of the state's farm sector may have little bearing on the prominence of rural communities. For example, California, Texas, Florida, and Illinois, and to a lesser extent Minnesota and Wisconsin, are all large contributors to national farm production but are predominately urban states. (See Appendix Table 1 for a breakdown, by state, of the distribution of the U.S. population.) This contrast may suggest significant state pressures to balance urban and farm interests. Allocation of water in western states, particularly California, is one of the most prominent examples of competing farm and urban needs. Establishing animal production facilities near urban and suburban areas constitutes another area of friction between urban population and the farming sector in some states.
- Some states have significant rural populations—like Delaware, Maine, and Vermont—but small farm sectors that contribute minimally to either national or state farm output. For these states, rural community and economic development issues may diverge significantly from farm issues. A few states, such as Idaho, Iowa, South Dakota, North Dakota, and Montana, have both large rural populations and large farm sectors that contribute significantly to the state economy (Table 2-8 and see Appendix Table 1). In these states, rural and farm issues may still be closely intertwined.
- At the same time that the agricultural concerns and priorities of local communities and individual states are changing, U.S. agriculture and its needs are increasingly shaped by international forces. For example, recent international accords like the North American Free Trade Agreement and those reached under the auspices of the General Agreement on Tariffs and Trade increase the integration of U.S. agriculture into global commodity and food markets and may limit the use of trade policies and subsidies to protect agriculture from international competition. In this environment, U.S. agriculture looks to other countries for new customers for its products and to science and technology for ways to stay ahead in intensely competitive markets. Also, while U.S. agriculture contributes, along with the agricultural industries of other nations, to today's abundant world food supplies, rapid world population growth leads many to stress the importance of sustaining and enhancing the productivity of the world's food-producing resources.

**TABLE 2-10** U.S. Employment on Farms and in Farm-Related Industries, by State, 1990

| State | Total Farm or Farm-Related Jobs | Percent of All State Employment | | | | | | |
|---|---|---|---|---|---|---|---|---|
| | | Farm Produc-tion | Agri-cultural Services | Agri-cultural Inputs | Agricultural Processing/ Marketing | Wholesale/ Retail Trade | Indirect Agri-business | Percent of All State Jobs |
| Alabama | 389,152 | 3.1 | 0.3 | 0.4 | 6.0 | 9.1 | 0.5 | 19.3 |
| Alaska | 51,425 | 0.2 | 5.6 | 0.1 | 1.8 | 8.9 | 0.1 | 16.6 |
| Arizona | 245,101 | 1.1 | 0.3 | 0.1 | 0.7 | 11.2 | 0.1 | 13.5 |
| Arkansas | 255,716 | 5.6 | 0.6 | 0.6 | 5.2 | 9.0 | 0.9 | 21.8 |
| California | 2,277,497 | 1.6 | 0.4 | 0.2 | 1.9 | 9.7 | 0.3 | 14.0 |
| Colorado | 289,135 | 2.2 | 0.2 | 0.2 | 1.7 | 10.2 | 0.2 | 14.7 |
| Connecticut | 222,699 | 0.4 | 0.2 | 0.1 | 1.0 | 9.2 | 0.3 | 11.2 |
| Delaware | 55,458 | 1.2 | 0.2 | 0.2 | 2.2 | 9.4 | 0.3 | 13.4 |
| Florida | 1,054,782 | 1.4 | 0.4 | 0.3 | 1.3 | 12.1 | 0.2 | 15.7 |
| Georgia | 676,391 | 2.0 | 0.3 | 0.3 | 5.1 | 10.1 | 0.7 | 18.6 |
| Hawaii | 114,573 | 2.1 | 0.1 | 0.1 | 2.1 | 12.2 | 0.1 | 16.6 |
| Idaho | 120,723 | 6.9 | 0.6 | 1.1 | 3.7 | 10.4 | 0.6 | 23.3 |
| Illinois | 910,990 | 1.7 | 0.2 | 0.6 | 1.8 | 9.6 | 0.6 | 14.4 |
| Indiana | 481,566 | 2.9 | 0.2 | 0.4 | 1.7 | 10.3 | 0.5 | 16.0 |
| Iowa | 399,904 | 8.3 | 0.4 | 1.9 | 3.8 | 10.1 | 0.5 | 25.0 |
| Kansas | 283,177 | 5.8 | 0.3 | 0.8 | 2.7 | 9.3 | 0.5 | 19.3 |
| Kentucky | 398,446 | 6.7 | 0.3 | 0.4 | 3.4 | 10.1 | 0.5 | 21.3 |
| Louisiana | 306,055 | 2.5 | 0.3 | 0.4 | 1.8 | 10.4 | 0.4 | 15.9 |
| Maine | 115,678 | 1.8 | 0.5 | 0.1 | 3.9 | 10.4 | 0.4 | 17.2 |
| Maryland | 336,347 | 0.9 | 0.3 | 0.1 | 1.3 | 9.9 | 0.3 | 12.7 |
| Massachusetts | 474,847 | 0.3 | 0.3 | 0.1 | 1.6 | 10.2 | 0.4 | 12.8 |
| Michigan | 654,206 | 1.7 | 0.2 | 0.2 | 1.0 | 10.6 | 0.3 | 14.0 |
| Minnesota | 458,945 | 4.5 | 0.2 | 0.6 | 2.1 | 9.7 | 0.4 | 17.4 |
| Mississippi | 246,016 | 4.8 | 0.4 | 0.6 | 5.5 | 9.4 | 0.6 | 21.2 |
| Missouri | 519,557 | 4.4 | 0.3 | 0.5 | 2.6 | 9.3 | 0.4 | 17.5 |
| Montana | 85,951 | 7.5 | 0.5 | 0.6 | 1.0 | 10.7 | 0.5 | 20.7 |
| Nebraska | 220,329 | 7.3 | 0.4 | 1.6 | 3.6 | 9.9 | 0.2 | 23.0 |
| Nevada | 75,596 | 0.7 | 0.2 | 0.1 | 0.3 | 9.3 | 0.1 | 10.7 |
| New Hampshire | 88,704 | 0.7 | 0.3 | 0.1 | 1.3 | 11.1 | 0.3 | 13.9 |
| New Jersey | 553,946 | 0.3 | 0.2 | 0.1 | 2.0 | 9.6 | 0.5 | 12.7 |
| New Mexico | 104,949 | 2.6 | 0.2 | 0.2 | 0.9 | 10.7 | 0.3 | 14.8 |
| New York | 1,232,735 | 0.7 | 0.2 | 0.1 | 1.9 | 9.5 | 0.3 | 12.6 |
| North Carolina | 841,852 | 2.5 | 0.2 | 0.3 | 8.6 | 9.8 | 0.6 | 22.0 |
| North Dakota | 93,489 | 11.7 | 0.3 | 1.4 | 2.2 | 10.1 | 0.1 | 25.7 |
| Ohio | 832,274 | 1.9 | 0.2 | 0.2 | 1.3 | 10.3 | 0.5 | 14.4 |
| Oklahoma | 274,284 | 5.1 | 0.3 | 0.3 | 1.8 | 9.7 | 0.1 | 17.2 |
| Oregon | 277,759 | 4.0 | 0.9 | 0.4 | 1.7 | 10.5 | 0.3 | 17.7 |
| Pennsylvania | 956,095 | 1.3 | 0.2 | 0.2 | 3.1 | 10.3 | 0.4 | 15.4 |
| Rhode Island | 74,301 | 0.3 | 0.2 | 0.1 | 2.2 | 10.7 | 0.4 | 13.8 |
| South Carolina | 380,472 | 2.0 | 0.2 | 0.2 | 7.1 | 10.0 | 0.7 | 20.2 |
| South Dakota | 96,820 | 11.0 | 0.3 | 0.9 | 2.8 | 10.0 | 0.2 | 25.2 |
| Tennessee | 532,984 | 4.2 | 0.2 | 0.3 | 4.6 | 9.8 | 0.6 | 19.7 |
| Texas | 1,325,087 | 2.5 | 0.2 | 0.3 | 1.8 | 10.2 | 0.3 | 15.2 |
| Utah | 127,277 | 2.2 | 0.2 | 0.2 | 1.7 | 10.1 | 0.4 | 14.6 |
| Vermont | 57,205 | 3.1 | 0.4 | 0.3 | 1.7 | 11.3 | 0.2 | 17.0 |
| Virginia | 537,111 | 1.7 | 0.3 | 0.2 | 3.2 | 9.2 | 0.4 | 14.9 |
| Washington | 444,110 | 3.0 | 0.8 | 0.3 | 1.6 | 10.4 | 0.3 | 16.4 |
| West Virginia | 118,841 | 3.2 | 0.3 | 0.4 | 1.5 | 10.4 | 0.3 | 16.0 |
| Wisconsin | 528,714 | 4.2 | 0.3 | 0.7 | 2.6 | 10.5 | 0.8 | 19.1 |
| Wyoming | 44,589 | 5.0 | 0.3 | 0.4 | 0.6 | 9.9 | 1.8 | 18.0 |
| Total | 21,285,425 | 2.3 | 0.3 | 0.3 | 2.4 | 10.0 | 0.4 | 15.7 |

NOTE: Farm and farm-related industries provided 25 percent or more of all jobs in Iowa, North Dakota, and South Dakota. Total percent of all state jobs may reflect rounding error.

SOURCE: Data were provided by the USDA Economic Research Service (ERS).

### ISSUES FOR DISCUSSION

- How have the dramatic changes in the U.S. economy—particularly in the role of farming—affected the interests of U.S. citizens in the services of the land grant colleges of agriculture? In particular, how are the interests of urban, suburban, and non-farm rural residents shaping the programs and priorities of colleges of agriculture?
- Although consumers have clearly benefited from agricultural science and technology, many perceive farmers to be the colleges' "traditional" clientele; but what type of farmer does today's and tomorrow's college serve? Are the needs and priorities of large commercial entities and vertically integrated operations the same as those of smaller, part-time, limited resource, or hobby farmers?
- Over time, states have become increasingly diverse with respect to the roles of farming, agribusiness, and rural communities and the way in which agricultural issues interact with other state issues. Are colleges of agriculture differently adapting their programs to the particular needs of their states?
- What is the role of the land grant colleges of agriculture in working with farmers and agricultural firms in adapting to increasingly open and competitive world markets? As global populations and food needs continue to grow, what is the role of U.S. colleges in contributing to the productivity and sustainability of agriculture world wide?

# SUGGESTED READINGS


Alston, Julian M., Philip G. Pardey, and Harold O. Carter, eds. *Valuing UC Agricultural Research and Extension.* Davis: University of California, 1994.

Caswell, Margriet, Keith O. Fuglie, and Cassandra A. Klotz. *Agricultural Biotechnology: An Economic Perspective*, AER No. 687. Washington, D.C.: Economic Research Service, U.S. Department of Agriculture, 1994.

Cochrane, Willard W. *The Development of American Agriculture: An Historical Analysis.* Minneapolis: University of Minnesota Press, 1979.

Dacquel, Laarni T., and Donald C. Dahmann. *Residents of Farms and Rural Areas: 1991*, U.S. Bureau of the Census, Current Population Reports Series P20, No. 472. Washington, D.C.: U.S. Government Printing Office, 1993.

Dahmann, Donald C., and Laarni T. Dacquel. *Residents of Farms and Rural Areas: 1990*, U.S. Bureau of the Census, Current Population Reports Series P-20, No. 457. Washington, D.C.: U.S. Government Printing Office, 1992.

Dodson, Charles B. *Profitability of Farm Businesses: A Regional, Farm Type, and Size Analysis*, Statistical Bulletin No. 884. Washington, D.C.: Economic Research Service, U.S. Department of Agriculture, 1994.

Frazão, Elizabeth. *Consumer Concerns About Nutrition: Opportunities for the Food Sector*, AIB No. 705. Washington, D.C.: Economic Research Service, U.S. Department of Agriculture, 1994.

Hoppe, Robert A. *Farming Operations and Households in Farming Areas: A Closer Look*, AER No. 685. Washington, D.C.: Economic Research Service, U.S. Department of Agriculture, 1994.

Leones, Julie, Gerald Schulter, and George Goldman. "Redefining agriculture in interindustry analysis." American Journal of Agricultural Economics 76(December 1994):1123-1129.

Kennedy, Eileen, and Howarth E. Boui. *Linkages Between Agriculture and Nutrition: Implications for Policy and Research.* Washington, D.C.: International Food Policy Research Institute, 1993.

National Research Council. *Alternative Agriculture.* Washington, D.C.: National Academy Press, 1989.

National Research Council. *Pesticides in the Diets of Infants and Children.* Washington, D.C.: National Academy Press, 1993.

Peterson, R. Neal, and Nora L. Brooks. *The Changing Concentration of U.S. Agricultural Production During the 20th Century*, AIB No. 671. Washington, D.C.: Economic Research Service, U.S. Department of Agriculture, 1993.

Pinstrup-Andersen, Per. *World Food Trends and Future Food Security* . Washington, D.C.: International Food Policy Research Institute, 1994.

Sommer, Judith E., and Fred K. Hines. *Diversity in U.S. Agriculture: A New Delineation by Farming Characteristics*, AER No. 646. Washington, D.C.: Economic Research Service, U.S. Department of Agriculture, 1991.

Vandeman, Ann, Jorge Fernandez-Cornejo, Sharon Jans, and Biing-Hwan Lin. *Adoption of Integrated Pest Management in U.S. Agriculture*, AIB No. 707. Washington, D.C.: Economic Research Service, U.S. Department of Agriculture, 1994.

# 3

# THE COLLEGES OF AGRICULTURE: Academic Programs and Demographics of Students and Graduates

*This chapter introduces the first of the land grant colleges' functions, that of academic instruction. It draws on data collected, compiled, and maintained by the Food and Agricultural Education Information System (FAEIS), which is supported by the U.S. Department of Agriculture and managed by Texas A&M University, to report trends in student enrollment and graduates, student demographics, and the types of agriculture-related degrees of most interest to students at different stages of training. This chapter also utilizes the results of a survey of Ph.D. scientists, conducted by the National Research Council's Office of Scientific and Engineering Personnel, to compare employment and demographic characteristics of agricultural scientists to those of all other scientists.*

- When the Morrill Act of 1862 was enacted, the framers undoubtedly intended that a teacher would be hired by the college to teach practical skills in farming and the mechanical arts. In keeping with the tradition of the European educational system, that would mean hiring a "professor of agriculture"—someone with extensive, broad-based knowledge of the subject—to lead a cadre of support staff to help carry out the teaching function. Although this model had worked well in such fields as theology, philosophy, law, medicine, and the classics, it did not fit the needs of U.S. agriculture. The simplest reason for why it did not is that the expertise needed for animal-based agriculture is very different than the expertise needed for plant-based agriculture.
- In 1862 there were virtually no persons trained either in agriculture or in the sciences relating to agriculture; and so the colleges, often by trial and error, had to develop their own faculty members, sometimes by recruiting highly skilled farmers. Although these farmers had expertise in animal or crop production, they did not have the requisite expertise in both. Consequently, from the outset, land grant colleges began to move from the generalist in agriculture to the specialist in agriculture. With this move came the need for more involvement from more people.

- By the end of the 19th century most colleges of agriculture already had several departments in their administrative structure. In the 20th century the degree of specialization, and the number of specialists, increased exponentially as the knowledge base expanded (in large part as the result of the Hatch Act, which established a research function in the agricultural experiment station). It became necessary to teach increasingly specialized courses not only to undergraduate students, but most particularly to the graduate students who would soon be in a position to make practical use of this new knowledge base both in the field and as teachers and researchers. Consequently, there was an explosion of curricula, majors, and options—and people—with a high degree of specialization (see box copy, p. 36).
- The trend toward increasing specialization, with its consequent increase in organizational structuring to accommodate curricula, majors, and options, continued until recently when it was recognized that issues facing society are exceedingly complex and require interdisciplinary teams to work on solutions. That realization is causing reorganization within many colleges of agriculture and the creation of interdisciplinary programs, centers, and institutes. Examples of this are degree programs in environmental sciences, centers for sustainable agriculture, and centers for biotechnology. These centers can exist as hard-wall entities (where faculty is brought together in one building), but many are instituted as soft-wall centers (where faculty is disbursed in various academic departments but come together for programmatic needs). Integration of such disciplines as animal science, horticulture, agronomy, plant pathology, entomology, natural resources, agricultural economics, and rural sociology, among others, may play an important role in the future of U.S. agriculture and of the land grant system.
- Higher education in agriculture, food, and natural resource sciences is supported by USDA grants. The major education grants administered by the USDA Office of Higher Education include: the higher education challenge grants program ($4 million in 1995); the higher education multicultural scholars program ($1 million in 1995); the 1890 institution capacity building grants program (about $9 million in 1995); and the USDA food and agricultural sciences national needs graduate fellowship grants program ($3.5 million in 1995) (National Association of State Universities and Land-Grant Colleges, 1995).

Land grant colleges of agriculture account for about 1 percent of all students enrolled at public institutions of higher education, but for higher percentages at land grant universities.

## ACADEMIC SPECIALIZATIONS IN AGRICULTURE, FORESTRY, AND RENEWABLE NATURAL RESOURCES

### General Agriculture

Agricultural sciences, general
Agricultural sciences, other

### Animal Sciences

Animal sciences, general
Animal breeding and genetics
Animal health
Animal nutrition
Animal physiology
Dairy science
Livestock
Poultry sciences
Pre-veterinary medicine
Veterinary medicine
Embryology
Endocrinology
Animal pathology
Animal pharmacology
Animal sciences, other

### Plant Sciences

Plant sciences, general
Agronomy
Horticulture science
Ornamental horticulture
Plant breeding and genetics
Plant pathology (applied)
Plant physiology
Plant protection (integrated pest management)
Turf management science
Landscape architecture
Plant pharmacology
Plant sciences, other

### Soil Sciences

Soil sciences, general
Soil chemistry
Soil conservation
Soil management and fertility
Soil microbiology
Soil physics
Soil sciences, other

### Agricultural Business and Management

Agricultural business and management, general
Agricultural business
Agricultural economics
Farm and ranch management
Agricultural business and management, other

### Education, Communication, and Social Sciences

International agriculture, general
Rural sociology, general
Agricultural communications/journalism, general
Extension education
Education, communication, social sciences, other

### Natural Resources

Fisheries science
Range management
Renewable natural resources conservation, general
Environmental science/studies
Natural resources management and policy
Natural resources law enforcement and protective services
Wildlife and wildlands management
Parks, recreation, and leisure studies
Parks, recreation and leisure facilities management
Water resources
Natural resources, other

| | |
|---|---|
| **Forest Sciences** | |
| Forest harvesting and production | Forest management |
| Forest products technology | Forest mensuration |
| Logging/timber harvesting | Urban forestry |
| Forestry, general | Wood science |
| Forest sciences | Pulp and paper technology |
| Forest biology | Forest soils |
| Forest engineering | Forestry and related sciences, other |
| Forest hydrology | |
| **Agricultural Engineering/Mechanization** | |
| Agricultural mechanics | Agricultural engineering |
| Agricultural mechanization | |
| **Food Science/Human Nutrition** | |
| Food sciences, general | Food technology |
| Dairy processing | Nutritional sciences |
| Food distribution | Food science/human nutrition, other |
| Food engineering | |
| Food packaging | |
| **Related Biological/Physical Science** | |
| Biology, general | Biometrics and biostatistics |
| Biochemistry and biophysics | Parasitology |
| Botany | Entomology |
| Mycology | Climatology/meteorology |
| Microbiology/bacteriology | Biological/physical science, other. |

SOURCE: Food and Agricultural Education Information System. 1994. Fall 1993 Enrollment in Agriculture, Renewable Natural Resources and Forestry: A Combined Report. College Station, Texas: Texas A&M University.

## ENROLLMENT: LAND GRANT VERSUS NON-LAND GRANT

- In the fall of 1991 there were approximately 10 million undergraduates and 1 million graduate students (both full and part time) enrolled in U.S. public institutions of higher education (U.S. Department of Education, 1993). Between 80 and 85 thousand of these undergraduates (<1 percent) and about 22 thousand graduate students ($\approx$ 2 percent) were enrolled at land grant colleges of agriculture (Table 3-1).[1]

[1] To provide information representative of 100 percent response, extrapolation was accomplished by applying the percent change by major area and degree level observed in institutions reporting in both 1992 and 1993 to responses from institutions that did not respond in either year. This process requires response from each institution in at least one year.

TABLE 3-1 Fall Enrollment, by Degree Program, at Land Grant Colleges of Agriculture, 1984–1993

| | Year | | | | |
|---|---|---|---|---|---|
| Degree Program | 1984 | 1986 | 1990 | 1992 | 1993 |
| Associate's | | | | | |
| Agricultural | 3,170 | 3,757 | 3,929 | 3,633 | 3,451 |
| Other | 110 | 483 | 518 | 70 | 59 |
| Subtotal | 3,280 | 4,240 | 4,447 | 3,703 | 3,510 |
| Bachelor's | | | | | |
| Agricultural | 71,241 | 63,232 | 66,390 | 71,706 | 78,192 |
| Other | 10,161 | 8,730 | 9,882 | 9,398 | 6,513 |
| Subtotal | 81,402 | 71,961 | 76,272 | 81,104 | 84,706 |
| Master's | | | | | |
| Agricultural | 12,831 | 13,002 | 10,873 | 11,082 | 11,751 |
| Other | 732 | 755 | 712 | 765 | 592 |
| Subtotal | 13,563 | 13,758 | 11,585 | 11,847 | 12,343 |
| Doctorate | | | | | |
| Agricultural | 9,197 | 9,412 | 9,990 | 9,753 | 10,032 |
| Other | 254 | 39 | 108 | 205 | 159 |
| Subtotal | 9,451 | 9,451 | 10,098 | 9,957 | 10,190 |

NOTE: To provide information representative of 100 percent response, extrapolation is based on applying the percent change by major area and degree level observed in institutions reporting in both 1992 and 1993 to responses from institutions that did not respond in either year. This process requires response from each institution in at least 1 year. Agricultural programs include natural resources, and forestry sciences; "other" includes any nonagricultural program such as chemistry, geography, geology, home economics, psychology, sociology, statistics, etc.
SOURCE: Data are from the Food and Agricultural Education Information System (FAEIS).

- In contrast to these aggregate data, at many land grant universities enrollment in colleges of agriculture is a larger percentage of university-wide or campus-wide enrollment. For example, at the University of California, Davis—the most "agricultural" of the University of California campuses—in the early 1990s about 25 percent of campus enrollment was in the college of agriculture. At North Carolina State University, enrollment in the college of agriculture was about 13 percent of university-wide enrollment; at Mississippi State it was about 10 percent. At the University of Illinois and the Pennsylvania State University, college of agriculture enrollment was 6 to 7 percent of university enrollment (U.S. Department of Education, 1993).
- As noted in Chapter 1, some non-land grant colleges and universities also have agriculture schools or colleges. In the fall of 1993, nationwide enrollment in all colleges of agriculture, renewable natural resources, and forestry comprised more than 137,000 students, 80 percent of whom were enrolled in the land grant colleges of agriculture (Food and Agricultural Education Information System, 1994; Table 3-1).
- Although total enrollment at public colleges and universities increased steadily from the mid-1980s through the early 1990s, land grant colleges of agriculture struggled to keep students (U.S. Department of Education, 1993). For example, enrollment in the colleges' bachelor's degree programs fell off sharply between 1984 and 1986 but recovered fully by 1992 (Table 3-1). Most colleges went on to report higher enrollment in 1993 than in 1992 (Table 3-2).

**TABLE 3-2** Fall Enrollment, by Region, at Land Grant Colleges of Agriculture, 1990 and 1992

| University | Undergraduate 1990 | Undergraduate 1992 | Graduate 1990 | Graduate 1992 |
|---|---|---|---|---|
| | *North-Central* | | | |
| Southern Illinois State U. | 700 | 788 | 98 | 104 |
| U. of Illinois | 1,815 | 1,897 | 595 | 629 |
| Purdue U. | 1,839 | 2,042 | 439 | 516 |
| Iowa State U. | 2,248 | 2,392 | 635 | 797 |
| Kansas State U. | 1,440 | 1,634 | 367 | 406 |
| Michigan State U. | 2,507 | 2,521 | 682 | 782 |
| U. of Minnesota | 747 | 1,253 | 507 | 517 |
| U. of Missouri | 1,648 | 1,846 | 417 | 411 |
| Lincoln U. | NR | 184 | NR | 0 |
| North Dakota State U. | NR | 738 | NR | 169 |
| U. of Nebraska | 1,139 | 1,344 | 478 | 366 |
| The Ohio State U. | 1,880 | 1,694 | 517 | 555 |
| South Dakota State U. | 1,077 | 1,440 | 190 | 212 |
| U. of Wisconsin, Madison | 1,889 | 2,025 | 1,227 | 1,256 |
| Regional total | 18,929 | 21,798 | 6,152 | 6,720 |
| | *Northeastern* | | | |
| U. of Connecticut | 421 | 550 | 144 | 161 |
| U. of Delaware | 627 | 605 | 61 | 105 |
| Delaware State U. | 64 | 82 | 0 | 0 |
| U. of Massachusetts | 1,872 | 2,192 | 408 | 428 |
| U. of Maryland | 830 | 795 | 267 | 251 |
| U. of Maryland, Eastern Shore | 92 | 147 | 18 | 12 |
| U. of Maine, Orono | 636 | 651 | 0 | 49 |
| U. of New Hampshire | 1,158 | 1,587 | 186 | 184 |
| Rutgers—The State U. of New Jersey | 2,862 | 2,908 | 560 | 604 |
| Cornell U. | 3,029 | 3,081 | 1,085 | 1,128 |
| The Pennsylvania State U. | 1,912 | 2,194 | 358 | 415 |
| U. of Rhode Island | 664 | 808 | 190 | 0 |
| U. of Vermont | 787 | 724 | 83 | 84 |
| West Virginia U. | 1,170 | 1,319 | 181 | 190 |
| Regional total | 16,124 | 17,643 | 3,541 | 3,611 |
| | *Southern* | | | |
| Auburn U. | 512 | 645 | 248 | 282 |
| Alabama A&M U. | 339 | 357 | 109 | 167 |
| U. of Arkansas, Pine Bluff | 340 | 196 | 0 | 1 |
| U. of Arkansas, Fayetteville | 795 | 868 | 285 | 270 |
| U. of Florida | 1,049 | 1,377 | 695 | 457 |
| Florida A&M U. | NR | 146 | NR | 21 |
| U. of Georgia | 913 | 1,031 | 241 | 247 |
| U. of Guam | 13 | 14 | 0 | 0 |
| U. of Kentucky | 745 | 890 | 276 | 259 |
| Louisiana State U. | 1,028 | 1,008 | 564 | 463 |
| Southern U. | NR | 287 | NR | 0 |
| Mississippi State U. | 1,070 | 1,104 | 313 | 354 |
| Alcorn State U. | 100 | 150 | 0 | 33 |
| North Carolina State U. | 2,677 | 3,034 | 707 | 742 |
| North Carolina A&T U. | 239 | NR | 61 | NR |
| Oklahoma State U. | 1,187 | 1,327 | 297 | 356 |
| Langston U. | 28 | 34 | 0 | 0 |
| U. of Puerto Rico | 552 | 577 | 76 | 97 |
| Clemson U. | 541 | 821 | 232 | 309 |

| | Undergraduate | | Graduate | |
|---|---|---|---|---|
| University | 1990 | 1992 | 1990 | 1992 |
| South Carolina State U. | 83 | 94 | 6 | 17 |
| U. of Tennessee | 759 | 836 | 229 | 224 |
| Tennessee State U. | NR | 174 | NR | 17 |
| Texas A&M U. | 3,671 | 3,927 | 1,241 | 1,254 |
| Texas Tech U. | 1,151 | 1,155 | 157 | 151 |
| Prairie View A&M U. | 219 | 134 | 64 | 48 |
| Virginia Polytechnic Institute and State U. | 1,199 | 1,465 | 466 | 461 |
| Virginia State U. | 278 | 162 | 0 | 0 |
| U. of the Virgin Islands | 72 | 0 | NR | NR |
| Regional total | 19,560 | 21,813 | 6,267 | 6,230 |
| *Western* | | | | |
| U. of Alaska | 66 | 78 | 27 | 25 |
| U. of Arizona | 1,694 | 1,644 | 509 | 479 |
| Arizona State U. | 208 | 422 | 36 | 67 |
| U. of California, Berkeley | 767 | NR | 342 | NR |
| U. of California, Davis | 5,279 | 5,026 | 1,309 | 1,098 |
| U. of California, Riverside | 1,104 | NR | 306 | NR |
| Colorado State U. | 760 | 1,000 | 192 | 190 |
| U. of Hawaii | 388 | 397 | 216 | 182 |
| U. of Idaho | 512 | 179 | 212 | 187 |
| College of Micronesia | NR | 7 | NR | 0 |
| Montana State U. | 600 | 535 | 110 | 111 |
| U. of Nevada, Reno | 295 | 332 | 78 | 61 |
| New Mexico State U. | 1,087 | 1,128 | 185 | 230 |
| Oregon State U. | 724 | 790 | 431 | 406 |
| American Samoa Community College | NR | 0 | NR | 0 |
| Utah State U. | 519 | 613 | 133 | 134 |
| Washington State U. | 1,078 | 1,365 | 335 | 282 |
| U. of Wyoming | 598 | 635 | 142 | 160 |
| Regional total | 15,679 | 14,151 | 4,563 | 3,612 |
| Total | 70,292 | 75,405 | 20,523 | 20,173 |

NOTE: NR, no response.
SOURCE: Data are from the Food and Agricultural Education Information System (FAEIS).

- Enrollment in doctorate programs increased steadily over the same period (Table 3-1). Nonetheless, graduate student enrollment in agricultural sciences has been losing ground versus other areas of science. In 1981 graduate enrollment in agricultural sciences (not including agricultural economics or agricultural engineering) accounted for 4 percent of graduate enrollment in all sciences; by 1991 agricultural science's share had dropped to 3 percent. Graduate enrollment in agricultural engineering also declined as a percent of enrollment in all engineering during this same period (U.S. Department of Education, 1993).
- USDA has divided the states according to four major geographic regions—northeastern, north-central, southern, and western (Figure 3-1). The largest percentages of land grant college of agriculture students are in the north-central region, the "farm belt," and in the southern region. Each of the four regions, however, contributes to total national enrollment at colleges of agriculture. In 1992 no region contributed less than 17 percent of either graduate or undergraduate students (Figure 3-2).

Enrollment at land grant colleges of agriculture is concentrated at a few large institutions.

- In 1992, 39 percent of reported enrollment at the undergraduate level was concentrated in ten land grant colleges of agriculture (Table 3-3); four of these are in the northeastern region. More than 40 percent of graduate student enrollment was concentrated in a slightly different list of ten land grant colleges (Table 3-4). Overall, less than 15 percent of the schools enroll 40 percent of the land grant college of agriculture students.
- In 1993 women composed 37.5 percent of undergraduates and 35 percent of graduate students in agricultural programs at land grant colleges of agriculture. In contrast, at all U.S. institutions of higher education, more than one-half of both undergraduate and graduate students are women (U.S. Department of Education, 1993). Women are significantly better represented in the agriculture colleges' "other" programs, which include principally home economics but also chemistry, geology, geography, psychology, sociology, statistics, etc. (Table 3-5).

**FIGURE 3-1**
*The four geographic regions of the United States, as determined by USDA.*

Women and ethnic minorities constitute smaller percentages of enrollment at land grant colleges of agriculture than at all institutions of higher education.

- Ethnic minorities—including African-Americans, Asian-Americans, Hispanic-Americans, and Native Americans—represent a small, though recently growing, percentage of land grant college of agriculture enrollment—about 10 percent in 1993 versus 5 percent in 1984 (Table 3-5). About 20 percent of ethnic minorities pursuing bachelor degrees attend the 1890 institutions (FAEIS, 1994). At all U.S. institutions of higher education, ethnic minorities account for slightly more than 20 percent of undergraduates and about 14 percent of graduate students (U.S. Department of Education, 1993).
- Students who are not U.S. citizens make a significant contribution to enrollment at the colleges of agriculture, but mostly at the graduate level. In 1993 (the only year for which data is currently available) more than one-fourth of graduate students were citizens of other countries, suggesting that colleges of agriculture may contribute significantly to the development of the human capital of other nations (Table 3-5).

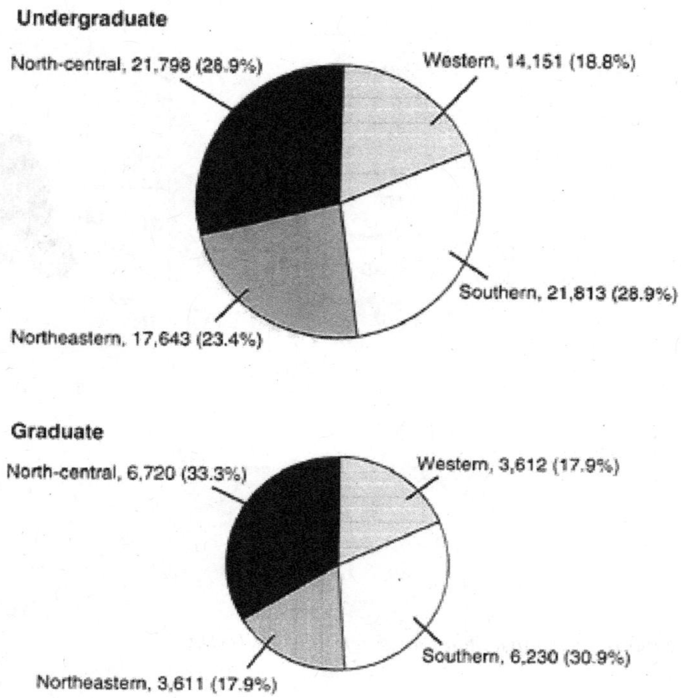


**FIGURE 3-2**

*The regional breakdown of undergraduate enrollment at land grant colleges of agriculture in the fall of 1992 shows nearly 60 percent of undergraduate enrollment evenly distributed between the north-central and southern regions. At the graduate level, the north-central and southern regions share 64 percent of the total enrollment.*

**TABLE 3-3** The Ten Land Grant Colleges of Agriculture Leading in Total College of Agriculture Undergraduate Enrollment, Fall 1992

| | Including Other Programs and Excluding Separately Administered Schools and Colleges of Forestry | | Excluding Other Programs and Separately Administered Schools and Colleges of Forestry | | Including Separately Administered Schools and Colleges of Forestry, Excluding Other Programs | |
|---|---|---|---|---|---|---|
| University | Number | Percent[a] | Number | Percent[a] | Number | Percent[a] |
| U. California, Davis | 5,026 | 7 | 3,552 | 5 | 3,552 | 5 |
| Texas A&M U. | 3,927 | 5 | 3,927 | 6 | 3,927 | 6 |
| Cornell U. | 3,081 | 4 | 3,081 | 5 | 3,081 | 4 |
| North Carolina State U. | 3,034 | 4 | 2,374 | 4 | 3,022 | 4 |
| Rutgers—The State U. of New Jersey | 2,908 | 4 | 1,555 | 2 | 1,555 | 2 |
| Michigan State U. | 2,521 | 3 | 2,141 | 3 | 2,141 | 3 |
| Iowa State U. | 2,392 | 3 | 2,392 | 4 | 2,392 | 3 |
| Pennsylvania State U. | 2,194 | 3 | 2,194 | 3 | 2,194 | 3 |
| U. of Massachusetts | 2,192 | 3 | 1,406 | 2 | 1,406 | 2 |
| Purdue U. | 2,042 | 3 | 2,042 | 3 | | |
| Subtotal | 29,317 | 39 | 24,664 | 37 | 25,312 | 36 |
| Total[b] | 75,850 | NA | 66,452 | NA | 71,270 | NA |

NOTE: "Other" programs include any nonagricultural programs offered by the college of agriculture such as home economics, statistics, sociology, chemistry, geology, geography, psychology, etc.

[a]Percent of all undergraduate students at all reporting colleges of agriculture.
[b]Total of undergraduates at all reporting colleges of agriculture.

SOURCE: S. S. Whatley, Project Coordinator, Food and Agricultural Education Information System, 1994, personal communication.

**TABLE 3-4** The Ten Land Grant Colleges of Agriculture Leading in Total College of Agriculture Graduate Enrollment, Fall 1992

| | Including Other Programs and Excluding Separately Administered Schools and Colleges of Forestry | | Excluding Other Programs and Separately Administered Schools and Colleges of Forestry | | Including Separately Administered Schools amd Colleges of Forestry and Excluding Other Programs | |
|---|---|---|---|---|---|---|
| University | Number | Percent[a] | Number | Percent[a] | Number | Percent[a] |
| U. of Wisconsin | 1,256 | 6 | 1,242 | 7 | 1,242 | 6 |
| Texas A&M U. | 1,254 | 6 | 1,254 | 7 | 1,254 | 6 |
| Cornell U. | 1,128 | 6 | 1,128 | 6 | 1,128 | 5 |
| U. of California, Davis | 1,098 | 5 | 994 | 5 | 994 | 5 |
| Iowa State U. | 797 | 4 | 797 | 4 | 979 | 5 |
| Michigan State U. | 782 | 4 | 774 | 4 | 774 | 4 |
| North Carolina State U. | 742 | 4 | 742 | 4 | 922 | 4 |
| U. of Illinois | 557 | 3 | 557 | 3 | 557 | 3 |
| Rutgers—The State U. of New Jersey | 604 | 3 | 392 | 2 | 392 | 3 |
| The Ohio State U. | 555 | 3 | 555 | 3 | 554 | 3 |
| Subtotal | 8,845 | 44 | 8,479 | 44 | 8,659 | 42 |
| Total[b] | 20,173 | NA | 19,203 | NA | 20,745 | NA |

NOTE: "Other" programs include any nonagricultural programs offered by the college of agriculture such as home economics, statistics, sociology, chemistry, geology, geography, psychology, etc.

[a]Percent of all graduate students at all reporting colleges of agriculture.
[b]Total of graduate students at all reporting colleges of agriculture.

SOURCE: S. S. Whatley, Project Coordinator, Food and Agricultural Education Information System, 1994, personal communication.

TABLE 3-5 Fall Enrollment, by Degree Program, of Female, Ethnic Minority, and Foreign Students at Land Grant Colleges of Agriculture, 1984–1993

| | 1984 | | 1986 | | 1990 | |
|---|---|---|---|---|---|---|
| Degree Program | Females | Ethnic Minorities | Females | Ethnic Minorities | Females | Ethnic Minorities |
| Bachelor's | | | | | | |
| Agricultural | 22,836 (32.1) | 2,683 (3.8) | 22,144 (35.0) | 2,684 (4.2) | 24,491 (36.9) | 4,195 (6.3) |
| Other | 6,648 (65.4) | 1,073 (10.6) | 6,210 (71.1) | 982 (71.1) | 6,892 (69.7) | 1,439 (14.6) |
| Subtotal | 29,484 (36.2) | 3,756 (4.6) | 28,354 (39.4) | 3,666 (5.1) | 31,383 (41.1) | 5,634 (7.4) |
| Graduate | | | | | | |
| Agricultural | 5,949 (27.0) | 1,168 (5.3) | 6,347 (28.3) | 1,119 (5.0) | 6,726 (32.2) | 1,812 (8.7) |
| Other | 45 (46.6) | 40 (4.1) | 413 (52.0) | 44 (5.5) | 470 (57.3) | 127 (15.5) |
| Subtotal | 6,408 (27.8) | 1,208 (5.2) | 6,760 (29.1) | 1,163 (5.0) | 7,196 (33.2) | 1,939 (8.9) |

NOTE: Number in parentheses is the percent of females and ethnic minorities compared with total student enrollment in each degree program. Total student enrollment figures are reported in Table 3-1. "Other" includes any nonagricultural program offered by the college of agriculture such as chemistry, geography, geology, home economics, psychology, sociology, statistics, etc.
SOURCE: Data are from the Food and Agricultural Education Information System (FAEIS).

TABLE 3-6 Graduates in Agriculture, Food, and Natural Resources, from All Degree Programs, by Region and Institution Classification, 1992

| Region | 1862s | 1890s | Non-Land Grants | All Institutions |
|---|---|---|---|---|
| North-central | 6,325 | 16 | 1,106 | 7,447 |
| Northeastern | 5,071 | 27 | 521 | 5,619 |
| Southern | 5,565 | 432 | 1,728 | 7,725 |
| Western | 4,339 | 0 | 1,463 | 5,802 |
| Total | 21,300 | 475 | 4,818 | 26,593 |

SOURCE: Data are from the Food and Agricultural Education Information System (FAEIS).

- Since the inception of colleges of agriculture, the science of agriculture has taken on a myriad of specializations. Although land grant universities—and principally their colleges of agriculture—grant most U.S. degrees in agriculture, food, and natural resources, non-land grant schools also grant a significant number of agriculture-related degrees, particularly in the southern and western regions (Table 3-6; Figure 3-3).
- In 1992, of all students graduating from all degree programs in the food, agriculture, and natural resources disciplines, 73 percent received bachelor's degrees (Figure 3-3). Natural resources and agricultural business and management (especially agricultural economics) accounted for nearly 40 percent of all bachelor's degrees issued (Table 3-7). Bachelor's degrees in animal science were also popular, probably because of the higher demand for admission to veterinary medicine colleges. Non-land grant schools granted 21 percent of all the bachelor's degrees (Table 3-8).

| | 1992 | | 1993 | | |
|---|---|---|---|---|---|
| Degree Program | Females | Ethnic Minorities | Females | Ethnic Minorities | Foreign Students |
| Bachelor's | | | | | |
| Agricultural | 26,809 (37.4) | 5,462 (7.6) | 29,351 (37.5) | 8,350 (10.7) | 1,095 (1.4) |
| Other | 5,418 (57.7) | 1,184 (12.6) | 4,054 (62.2) | 90 (13.8) | 147 (2.3) |
| Subtotal | 32,227 (39.7) | 6,646 (8.2) | 33,405 (39.4) | 9,251 (10.9) | 1,242 (1.5) |
| Graduate | | | | | |
| Agricultural | 7,166 (34.4) | 1,792 (8.6) | 7,653 (35.1) | 2,069 (9.5) | 5,669 (26.0) |
| Other | 683 (70.4) | 100 (10.3) | 427 (56.9) | 82 (10.9) | 154 (20.5) |
| Subtotal | 7,849 (36.0) | 1,892 (8.7) | 8,080 (35.9) | 2,151 (9.5) | 5.823 (25.8) |

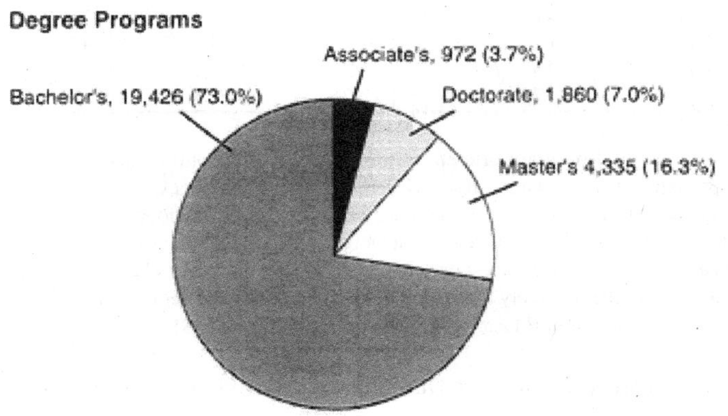


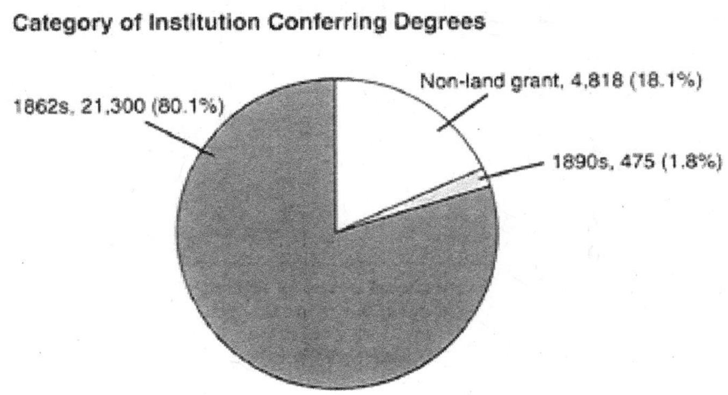


**FIGURE 3-3**
*Of all degrees conferred in agriculture, food, and natural resources in 1992, non-land grants conferred slightly more than 18 percent and bachelor degrees accounted for nearly three-quarters.*

TABLE 3-7 Number of Graduates in Various Agriculture Disciplines, by Degree Program, from All Institutions, 1992

| | Degree | | | | |
|---|---|---|---|---|---|
| Discipline[a] | Associate's | Bachelor's | Master's | Doctorate | All Degree Programs |
| General agriculture | 59 | 681 | 128 | 1 | 869 |
| Animal sciences | 183 | 2,653 | 443 | 230 | 3,509 |
| Plant sciences | 445 | 2,020 | 698 | 438 | 3,601 |
| Soil sciences | 12 | 116 | 102 | 76 | 306 |
| Agricultural business and management | 120 | 3,324 | 484 | 122 | 4,050 |
| Social sciences | 4 | 962 | 359 | 81 | 1,406 |
| Natural resources | 62 | 3,825 | 1,030 | 283 | 5,200 |
| Agricultural engineering/mechanization | 16 | 493 | 134 | 67 | 710 |
| Food sciences | 27 | 1,463 | 453 | 195 | 2,138 |
| Related sciences | 1 | 1,493 | 288 | 334 | 2,116 |
| Other | 43 | 2,399 | 216 | 33 | 2,691 |
| Total | 972 | 19,426 | 4,335 | 1,860 | 26,596 |

NOTE: "Other" includes any nonagricultural program offered by the college of agriculture such as chemistry, geography, geology, home economics, psychology, sociology, statistics, etc. "Natural resources" includes forest sciences.

[a] The relationship between the number of graduates and the number of practicing professionals in each academic specialization is unclear. For example, compared to the large membership of soil science societies, a relatively small number of graduates now receive soil science degrees, according to FAEIS. Reasons for the discrepancy may include changes in degree classification, participation by other agricultural scientists in soil science organizations, and the relatively higher popularity of soil science degrees in earlier years.

SOURCE: Data are from the Food and Agricultural Education Information System (FAEIS).

TABLE 3-8 Percent of Degrees in Various Agriculture Disciplines Conferred by Non-Land Grant Universities, by Degree Program, 1992

| | Degree Program | | | | |
|---|---|---|---|---|---|
| Discipline | Associate's | Bachelor's | Master's | Doctorate | All Degree Programs |
| General agriculture | 7 | 39 | 71 | 0 | 42 |
| Animal sciences | 9 | 19 | 7 | 2 | 16 |
| Plant sciences | 4 | 20 | 8 | 2 | 14 |
| Soil sciences | 0 | 36 | 8 | 0 | 16 |
| Agricultural business and management | 1 | 28 | 7 | 3 | 24 |
| Social sciences | 0 | 25 | 10 | 2 | 20 |
| Natural Resources | 0 | 25 | 32 | 17 | 26 |
| Agricultural engineering/ mechanization | 0 | 21 | 12 | 0 | 17 |
| Food Sciences | 22 | 13 | 2 | 0 | 9 |
| Related sciences | 0 | 1 | 1 | 0 | 1 |
| Other | 0 | 18 | 6 | 0 | 16 |
| Total | 5 | 21 | 14 | 4 | 18 |

NOTE: "Other" includes any nonagricultural program offered by the college of agriculture such as chemistry, geography, geology, home economics, psychology, sociology, statistics, etc. "Natural resources" includes forest sciences.

SOURCE: Data are from the Food and Agricultural Education Information System (FAEIS).

More than 80 percent of graduates in agriculture, food, and natural resources receive their degrees at land grant universities, but non-land grant schools are important in some fields.

- Master's degrees in natural resources accounted for one-quarter of the master's degrees issued (Table 3-7). The high number of master's degrees in natural resources may be because a master's degree is now forestry's first-level professional degree. In 1992 non-land grant schools granted about one-third of the natural resource degrees at the master's level (Table 3-8).
- Students in doctorate programs have a different orientation; they are most numerous in plant sciences and "related" sciences such as botany and entomology. Natural resources, animal sciences, and food sciences are the next most populated disciplines at the doctorate level (Table 3-7). Non-land grant schools conferred only a small share of these doctorate degrees; in other words, producing Ph.D. agricultural scientists is the domain of the land grant colleges (Table 3-8).
- From all degree programs, less than 5 percent of all graduates in 1992 received degrees labeled "general agriculture" or soil science (Table 3-7). More than 40 percent of the "general agriculture" degrees (which are mostly undergraduate degrees) were issued by non-land grant institutions (Table 3-8).
- There is some regional specialization in academic fields, particularly at the doctorate level. For example, doctorate degrees in food science are a higher percentage in the northeast than in other parts of the country; while more doctorate degrees in natural resources are conferred in the west (Figure 3-4). "Related science" degrees are most significant (as a percent of all degrees) in the northeast and least significant in the south (see Appendix Table 2).
- Across the country, colleges of agriculture offer many of the same agricultural science or other agriculture-related degree programs. For example, in 1991-1992 in the land grant system, there were 92 bachelor's degree programs in animal science and 50 at the doctorate level (Table 3-9). In plant sciences in that same academic year, there were 129 undergraduate degree programs including 35 in agronomy and 38 in horticulture (Table 3-9).
- There was a slight decline between 1985 and 1992 in the number of degree programs in most agricultural and renewable natural resource specializations offered at land grant colleges of agriculture, particularly at undergraduate and master's degree levels (Figures 3-5 through 3-7).
- Using Table 3-9 to more closely examine specific areas of specialization, data indicate a decline in the number of programs in traditional agricultural specializations like dairy and poultry science. Some colleges offer general animal science degrees as well as more specific degrees in dairy or poultry science. In most cases, these more specialized programs are being incorporated into the more general programs. The two-period comparison in Table 3-9 also suggests that some expansion is occurring in basic science degree programs and some natural resource specializations like water resources and wildlife management.

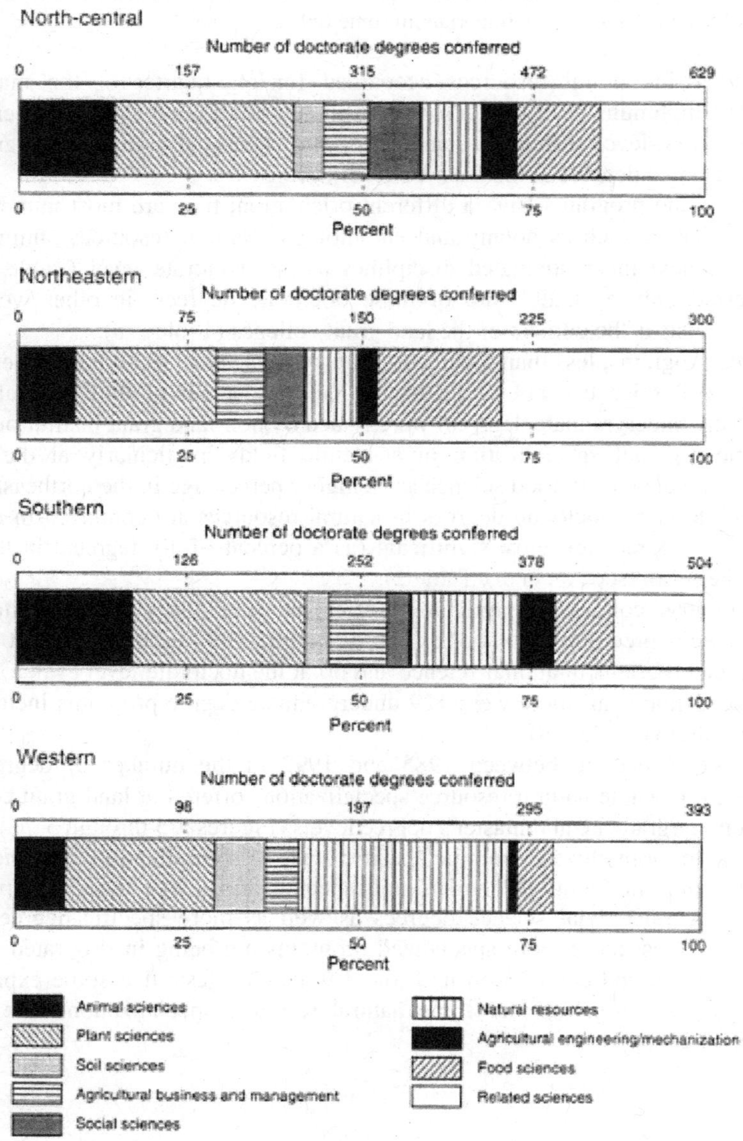


**FIGURE 3-4**

*The graph provides a regional breakdown of doctorate degrees conferred by land grant and non-land grant institutions in agriculture, food, and natural resources specializations in 1992.*

TABLE 3-9 Number of Degree Programs in Agricultural Science and Renewable Natural Resource Specializations at Land Grant Institutions, 1984–1985 and 1991–1992

| | Degree Program | | | | | |
|---|---|---|---|---|---|---|
| | Bachelor's | | Master's | | Doctorate | |
| Specialization | 1984–85 | 1991–92 | 1984–95 | 1991–92 | 1984–85 | 1991–92 |
| Agricultural business and management | | | | | | |
| Agricultural business/management, general | 12 | 11 | 2 | 1 | 1 | 1 |
| Agricultural business | 16 | 16 | 2 | 1 | 0 | 0 |
| Agricultural economics | 49 | 45 | 48 | 47 | 26 | 27 |
| Farm and ranch management | 3 | 3 | 1 | 1 | 0 | 0 |
| Agricultural business/management, other | 5 | 2 | 1 | 1 | 0 | 0 |
| Agricultural products/processing, general | 0 | 1 | 1 | 2 | 1 | 2 |
| Nonfood products | 0 | 0 | 0 | 0 | 0 | 0 |
| Agricultural products/processing, other | 0 | 1 | 0 | 0 | 0 | 0 |
| Floristry farm and garden supp, gen other | 0 | 0 | 0 | 0 | 0 | 0 |
| Total | 90 | 84 | 63 | 60 | 31 | 36 |
| Agricultural mechanics | | | | | | |
| Agricultural mechanics, general | 17 | 15 | 4 | 3 | 0 | 0 |
| Agricultural elect/power/controls | 0 | 0 | 0 | 0 | 0 | 0 |
| Agricultural mech/const/maint skills | 0 | 0 | 0 | 0 | 0 | 0 |
| Agricultural power machinery | 0 | 0 | 0 | 0 | 0 | 0 |
| Agricultural structures/equipment/facilities | 0 | 0 | 0 | 0 | 0 | 0 |
| Soil and water mechanical practices | 0 | 2 | 0 | 0 | 0 | 1 |
| Agricultural mechanics, other | 2 | 3 | 1 | 1 | 0 | 0 |
| Total | 19 | 20 | 5 | 4 | 0 | 1 |
| Agricultural services and supplies | | | | | | |
| Agricultural services and supplies, general | 1 | 0 | 0 | 0 | 0 | 0 |
| Agricultural services | 0 | 1 | 0 | 0 | 0 | 0 |
| Agricultural supplies marketing | 0 | 0 | 0 | 0 | 0 | 0 |
| Agricultural services and supplies, other | 0 | 0 | 0 | 0 | 0 | 0 |
| Total | 1 | 1 | 0 | 0 | 0 | 0 |
| International agriculture | 3 | 3 | 1 | 1 | 0 | 0 |
| Agricultural sciences | | | | | | |
| Agricultural production, general | 1 | 1 | 1 | 1 | 0 | 0 |
| Agricultural production, other | 0 | 0 | 0 | 1 | 0 | 0 |
| Agricultural sciences, general | 34 | 33 | 12 | 9 | 1 | 1 |
| Total | 35 | 34 | 13 | 11 | 1 | 1 |
| Animal sciences | | | | | | |
| Animal production | 1 | 0 | 1 | 1 | 0 | 0 |
| Aquaculture | 0 | 2 | 1 | 1 | 1 | 1 |
| Animal sciences, general | 56 | 57 | 48 | 45 | 28 | 31 |
| Animal breeding and genetics | 1 | 0 | 1 | 2 | 1 | 2 |
| Animal health | 1 | 2 | 2 | 2 | 1 | 1 |
| Animal nutrition | 1 | 0 | 3 | 1 | 2 | 2 |
| Animal physiology | 0 | 0 | 1 | 0 | 1 | 0 |
| Dairy science | 15 | 12 | 14 | 12 | 6 | 4 |
| Fisheries science | 4 | 0 | 3 | 0 | 3 | 0 |
| Livestock science | 0 | 0 | 0 | 0 | 0 | 0 |
| Poultry science | 15 | 13 | 15 | 7 | 8 | 6 |
| Animal sciences, other | 4 | 6 | 3 | 5 | 2 | 3 |
| Animal technology | 0 | 0 | 0 | 0 | 0 | 0 |
| Total | 98 | 92 | 92 | 76 | 53 | 50 |
| Food sciences | | | | | | |
| Food products | 1 | 2 | 0 | 0 | 0 | 0 |
| Food sciences, general | 27 | 33 | 26 | 31 | 19 | 25 |
| Dairy processing | 3 | 0 | 0 | 0 | 0 | 0 |
| Food distribution | 0 | 0 | 0 | 0 | 0 | 0 |
| Food engineering | 0 | 0 | 1 | 0 | 1 | 0 |
| Food packaging | 0 | 0 | 0 | 0 | 0 | 0 |
| Food technology | 5 | 0 | 2 | 0 | 2 | 0 |
| Food sciences, other | 2 | 0 | 2 | 0 | 2 | 0 |
| Food processing technology | 0 | 1 | 0 | 0 | 0 | 0 |
| Total | 38 | 36 | 31 | 31 | 24 | 25 |

| | Degree Program | | | | | |
|---|---|---|---|---|---|---|
| | Bachelor's | | Master's | | Doctorate | |
| Specialization | 1984–85 | 1991–92 | 1984–85 | 1991–92 | 1984–85 | 1991–92 |
| Plant sciences | | | | | | |
| Crop production | 0 | 1 | 1 | 2 | 1 | 1 |
| Horticulture, general | 8 | 10 | 6 | 7 | 3 | 4 |
| Arboriculture | 0 | 0 | 0 | 0 | 0 | 0 |
| Floriculture | 0 | 5 | 0 | 3 | 0 | 1 |
| Greenhouse operation and management | 0 | 0 | 0 | 0 | 0 | 0 |
| Landscaping | 2 | 5 | 0 | 0 | 0 | 0 |
| Nursery operation and management | 0 | 0 | 0 | 0 | 0 | 0 |
| Turf management | 0 | 1 | 0 | 0 | 0 | 0 |
| Horticulture, other | 1 | 1 | 0 | 0 | 1 | 1 |
| Plant sciences, general | 16 | 13 | 11 | 8 | 5 | 9 |
| Agronomy | 36 | 35 | 35 | 34 | 31 | 30 |
| Horticulture | 29 | 28 | 28 | 27 | 17 | 19 |
| Ornamental horticulture | 18 | 0 | 16 | 0 | 9 | 0 |
| Plant breeding and genetics | 0 | 0 | 4 | 0 | 5 | 0 |
| Plant pathology | 5 | 0 | 6 | 0 | 5 | 0 |
| Plant physiology | 0 | 0 | 0 | 0 | 1 | 0 |
| Plant protection (pest management) | 8 | 6 | 9 | 7 | 1 | 2 |
| Range management | 12 | 12 | 10 | 9 | 7 | 6 |
| Turf management science | 0 | 0 | 0 | 0 | 0 | 0 |
| Plant sciences, other | 3 | 6 | 3 | 3 | 1 | 2 |
| Plant genetics | 2 | 1 | 2 | 3 | 2 | 3 |
| Plant pathology | 6 | 5 | 23 | 29 | 21 | 26 |
| Plant physiology | 1 | 0 | 4 | 4 | 5 | 7 |
| Total | 147 | 129 | 158 | 136 | 115 | 111 |
| Soil sciences | | | | | | |
| Soil sciences, general | 23 | 22 | 19 | 27 | 14 | 18 |
| Soil chemistry | 0 | 0 | 1 | 0 | 0 | 0 |
| Soil conservation | 0 | 0 | 0 | 0 | 0 | 0 |
| Soil management and fertility | 0 | 0 | 2 | 0 | 2 | 0 |
| Soil microbiology | 0 | 0 | 0 | 0 | 1 | 0 |
| Soil physics | 0 | 0 | 0 | 0 | 1 | 0 |
| Soil sciences, other | 2 | 0 | 2 | 0 | 2 | 0 |
| Total | 25 | 22 | 24 | 27 | 20 | 18 |
| Agricultural sciences, other | | | | | | |
| Agribusiness and production, other | 9 | 7 | 4 | 2 | 1 | 1 |
| Agricultural sciences, other | 6 | 8 | 6 | 4 | 2 | 1 |
| Total | 15 | 15 | 10 | 6 | 3 | 2 |
| Renewable natural resources | | | | | | |
| Renewable natural resources, general | 25 | 21 | 17 | 15 | 6 | 6 |
| Renewable natural resources, other | 4 | 5 | 2 | 4 | 1 | 1 |
| Total | 29 | 26 | 19 | 19 | 7 | 7 |
| Conservation and regulation | | | | | | |
| Conservation and regulation, general | 1 | 3 | 0 | 0 | 0 | 0 |
| Conservation | 2 | 2 | 0 | 2 | 0 | 1 |
| Resources protection and regulation | 0 | 0 | 0 | 0 | 0 | 0 |
| Conservation and regulation, other | 1 | 1 | 0 | 0 | 0 | 0 |
| Total | 4 | 6 | 0 | 2 | 0 | 1 |
| Fishing and fisheries | | | | | | |
| Fishing and fisheries, general | 3 | 10 | 3 | 9 | 2 | 5 |
| Fisheries | 7 | 0 | 5 | 0 | 1 | 0 |
| Fishing and fisheries, other | 1 | 0 | 1 | 0 | 1 | 0 |
| Total | 11 | 10 | 9 | 9 | 4 | 5 |
| Forestry production and processing | | | | | | |
| Forestry production and processing, general | 4 | 4 | 3 | 3 | 4 | 4 |
| Forest production | 0 | 0 | 0 | 0 | 0 | 0 |
| Forest products utilization | 1 | 0 | 0 | 0 | 0 | 0 |
| Forest products processing technology | 2 | 1 | 0 | 0 | 0 | 0 |
| Pulp and paper production | 1 | 0 | 0 | 0 | 0 | 0 |
| Forestry production and processing, other | 2 | 2 | 0 | 0 | 0 | 0 |
| Total | 10 | 7 | 3 | 3 | 4 | 4 |

| | Degree Program | | | | | |
|---|---|---|---|---|---|---|
| | Bachelor's | | Master's | | Doctorate | |
| Specialization | 1984–85 | 1991–92 | 1984–85 | 1991–92 | 1984–85 | 1991–92 |
| Forestry and related sciences | | | | | | |
| Forestry and related products, general | 16 | 15 | 20 | 21 | 13 | 13 |
| Forestry | 7 | 5 | 5 | 6 | 1 | 3 |
| Forest biology | 2 | 0 | 0 | 0 | 0 | 0 |
| Forest engineering | 3 | 3 | 1 | 1 | 0 | 0 |
| Forest hydrology | 1 | 0 | 2 | 0 | 1 | 0 |
| Forest management | 9 | 8 | 4 | 3 | 3 | 2 |
| Forest mensuration | 0 | 0 | 0 | 0 | 0 | 0 |
| Urban forestry | 0 | 0 | 0 | 0 | 0 | 0 |
| Wood science | 8 | 7 | 6 | 4 | 0 | 2 |
| Forestry and related science, other | 3 | 4 | 1 | 1 | 0 | 0 |
| Total | 49 | 42 | 39 | 36 | 18 | 20 |
| Wildlife management | | | | | | |
| Game farm management | 0 | 0 | 0 | 0 | 0 | 0 |
| Wildlife management | 22 | 25 | 18 | 20 | 7 | 6 |
| Total | 22 | 25 | 18 | 20 | 7 | 6 |
| Agricultural engineering | 41 | 38 | 32 | 36 | 15 | 19 |
| Preveterinary | 4 | 4 | 0 | 0 | 0 | 0 |
| Veterinary medicine | | | | | | |
| Veterinary science (excludes D.V.M.) | 5 | 6 | 15 | 18 | 11 | 14 |
| Biology, general | 63 | 126 | 32 | 59 | 16 | 28 |
| Biochemistry and biophysics | 34 | 36 | 36 | 34 | 33 | 35 |
| Botany | | | | | | |
| Botany, general | 33 | 23 | 32 | 29 | 28 | 27 |
| Bacteriology | 4 | 2 | 3 | 1 | 2 | 0 |
| Mycology | 0 | 0 | 1 | 0 | 0 | 0 |
| Plant pharmacology | 0 | 0 | 0 | 0 | 0 | 0 |
| Total | 37 | 25 | 36 | 30 | 30 | 27 |
| Cell and molecular biology | | | | | | |
| Cell biology | 1 | 2 | 3 | 4 | 3 | 4 |
| Molecular biology | 2 | 4 | 2 | 3 | 3 | 4 |
| Total | 3 | 6 | 5 | 7 | 6 | 8 |
| Microbiology | 38 | 39 | 36 | 34 | 31 | 32 |
| Miscellaneous specialized areas, life sciences | | | | | | |
| Anatomy | 1 | 1 | 4 | 5 | 5 | 8 |
| Biometrics and biostatistics | 4 | 3 | 4 | 4 | 4 | 4 |
| Ecology | 8 | 8 | 14 | 12 | 11 | 11 |
| Embryology | 0 | 0 | 0 | 0 | 0 | 0 |
| Endocrinology | 0 | 0 | 2 | 1 | 1 | 3 |
| Histology | 0 | 0 | 0 | 0 | 0 | 0 |
| Marine biology | 2 | 3 | 4 | 3 | 2 | 2 |
| Parasitology | 0 | 10 | 2 | 14 | 0 | 18 |
| Toxicology | 1 | 0 | 5 | 9 | 8 | 8 |
| Total | 16 | 25 | 35 | 48 | 31 | 54 |
| Parks and recreation, general | 17 | 13 | 6 | 6 | 2 | 3 |
| Parks and recreation management | 24 | 22 | 8 | 8 | 1 | 2 |
| Water resources | 0 | 2 | 5 | 6 | 2 | 0 |

NOTE: The list of specializations in this table was derived from many sources that have similar programs, each of which has been titled slightly differently. Consequently, titles listed may seem repetitive or categories may seem fragmented. Data are provided in response to FAEIS's annual survey of enrollment (see below) and any specialization in which a degree was conferred is included in this listing.

SOURCE: FAEIS. 1994. Fall 1993 Enrollment in Agriculture, Renewable Natural Resources and Forestry: A Combined Report. College Station, Texas: Texas A&M University.

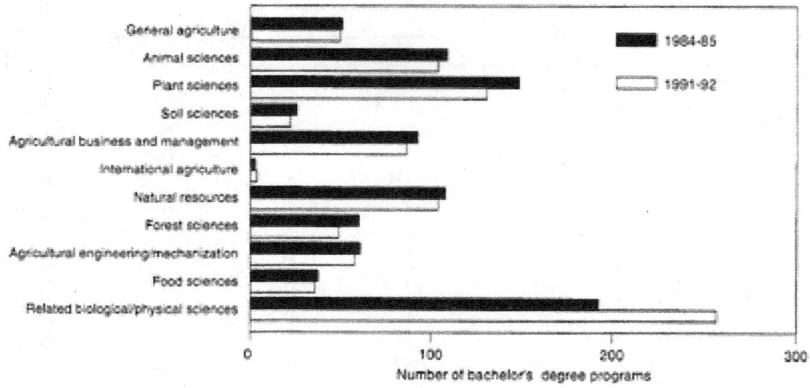


**FIGURE 3-5**

*The number of bachelor degree programs in most areas of agricultural science and renewable resources at land grant institutions declined slightly between 1984–1985 and 1991–1992, although the number of programs in general biology (genetic, cell, and molecular biology and microbiology) increased significantly.*

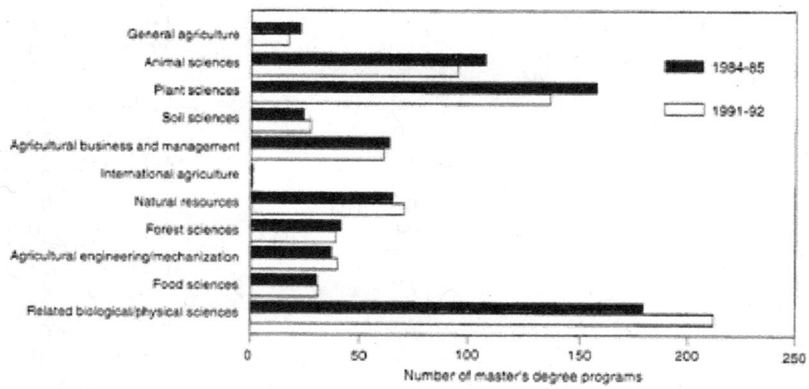


**FIGURE 3-6**

*At the master's degree level, between 1984–1985 and 1991–1992 the number of programs in soil science, natural resources, and agricultural engineering/mechanics at land-grant institutions increased only slightly while the number offering programs in general biology (genetic, cell, and molecular biology and microbiology) increased more.*

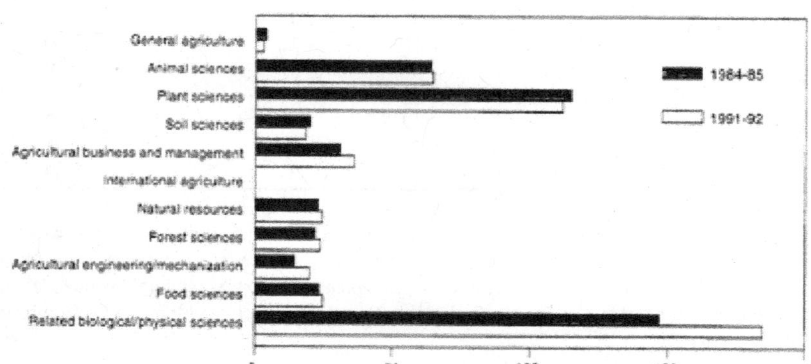


**FIGURE 3-7**

*Between 1984–1985 and 1991–1992 there were increases in more degree program areas at the doctorate level than at the bachelor or master's levels.*

## A DEMOGRAPHIC PROFILE OF AGRICULTURAL SCIENTISTS

- Today's and tomorrow's leaders of the land grant colleges of agriculture are its doctorate degree recipients. Its graduates at all degree levels have the potential to become leaders in industry, government, schools, and communities across the country—in fact, they often become the colleges' clientele. However, Ph.D. recipients are the ones most likely to join the colleges as faculty members and scientists, and advance to administrative positions at the colleges and universities.
- Most U.S. agricultural scientists received their doctorate degrees from land grant universities. Thus a review of demographic characteristics of Ph.D. agricultural scientists who received degrees from U.S. schools provides a fairly accurate profile of those who received doctorate degrees from land grant colleges of agriculture. The demographic data used here were collected by surveying a sample of Ph.D. scientists; the survey was administered by the National Research Council's Office of Science and Engineering Personnel.[2]
- Table 3-10 shows that the estimated number of agricultural scientists continues to grow but more slowly than the number of scientists in the life or natural sciences. In relative terms, agricultural science is losing human capital. Among agricultural scientists, the majority are plant and soil scientists, though the dominance of this group has declined from 58 percent in 1973 to 49 percent in 1991.
- Agricultural scientists with doctorate degrees, like their peers in other areas of science, are still most likely to go to work for an academic institution. However, agricultural scientists are more likely to be employed in government (because of the large intramural research program at the USDA) and less likely to be employed in the private sector than their natural science peers. For all scientists, including agricultural scientists, academia and government are becoming less important as employers, while industry is becoming more important (Table 3-11). (Food scientists are the agricultural scientists most likely to have private sector jobs.) Thus the private sector increasingly competes with universities for those graduates with leadership potential.
- Agricultural science is still clearly a more male-dominated discipline than other areas of science (Table 3-12). Within the agricultural science community, only in food sciences are women, at 22.5 percent, nearly as well represented as they are in science in general.

TABLE 3-10 Number of Employed Scientists Holding Doctorate Degrees, by Discipline, 1973, 1979, 1985, 1991

| | Year | | | |
|---|---|---|---|---|
| Discipline | 1973 | 1979 | 1985 | 1991 |
| Agricultural sciences | | | | |
| Agricultural economics | 314 | 993 | 1,835 | 2,200 |
| Animal sciences | 2,184 | 2,909 | 3,612 | 4,148 |
| Plant and soil sciences | 6,285 | 8,016 | 9,687 | 10,751 |
| Forestry and wildlife management | 1,346 | 2,211 | 2,510 | 2,811 |
| Food sciences | 653 | 915 | 1,430 | 1,944 |
| Subtotal | 10,782 | 15,044 (40) | 19,074 (27) | 21,854 (15) |
| Life sciences | 42,593 | 60,343 (42) | 81,226 (35) | 99,180 (22) |
| Natural sciences | 123,248 | 166,265 (35) | 208,431 (25) | 243,081 (17) |
| All sciences | 173,674 | 248,994 (43) | 323,056 (30) | 379,768 (18) |

NOTE: Number in parentheses is the percentage increase over previous year in table.

SOURCE: Data are from the National Research Council, Office of Scientific and Engineering Personnel's biennial surveys of doctorate recipients.

[2] Scientists answering the National Research Council survey identified their own scientific specializations. These may not match their academic degree specializations at graduation reported to FAEIS.

TABLE 3-11 Employment (percent) of Scientists Holding Doctorate Degrees, 1973, 1979, 1985, 1991

| | Year | | | |
|---|---|---|---|---|
| Employment Field | 1973 | 1979 | 1985 | 1991 |
| Education | | | | |
| Agricultural sciences | 62.8 | 57.8 | 57.0 | 54.5 |
| Life sciences | 68.2 | 66.5 | 62.2 | 59.1 |
| Natural sciences | 57.8 | 55.6 | 53.4 | 51.7 |
| All other sciences | 61.9 | 58.6 | 55.6 | 52.8 |
| Business/Industry[a] | | | | |
| Agricultural sciences | 15.0 | 19.8 | 24.6 | 25.0 |
| Life sciences | 12.4 | 13.3 | 19.5 | 22.2 |
| Natural sciences | 25.1 | 27.1 | 31.5 | 32.4 |
| All other sciences | 20.4 | 22.3 | 27.7 | 29.3 |
| Government[b] | | | | |
| Agricultural sciences | 20.7 | 19.5 | 15.7 | 16.3 |
| Life sciences | 12.4 | 11.6 | 10.1 | 10.4 |
| Natural sciences | 12.0 | 11.2 | 9.6 | 9.9 |
| All other sciences | 11.4 | 10.7 | 9.3 | 9.6 |
| Nonprofit Organizations | | | | |
| Agricultural sciences | 1.3 | 2.3 | 2.3 | 2.4 |
| Life sciences | 6.6 | 7.6 | 8.1 | 7.7 |
| Natural sciences | 4.8 | 5.4 | 5.3 | 5.3 |
| All other sciences | 6.0 | 7.5 | 7.0 | 7.2 |

[a] Includes self-employed.

[b] Federal, state, and local.

SOURCE: Data are from the National Research Council, Office of Scientific and Engineering Personnel's biennial surveys of doctorate recipients.

TABLE 3-12 Prevalent Demographic Characteristics (percent) of Employed Scientists Holding Doctorate Degrees, 1991

| | Characteristic | | | |
|---|---|---|---|---|
| Discipline | Male | White | Age 55 and Over | U.S. Citizen |
| Agricultural sciences | 90.0 | 86.3 | 19.9 | 92.8 |
| Agricultural economics | 91.4 | 85.6 | 5.8 | 88.6 |
| Animal sciences | 89.8 | 90.2 | 17.9 | 93.1 |
| Plant and soil sciences | 90.7 | 86.3 | 23.5 | 93.7 |
| Forestry and wildlife management | 92.5 | 93.9 | 22.9 | 95.5 |
| Food sciences | 80.5 | 65.1 | 16.9 | 87.9 |
| Life sciences | 73.6 | 87.8 | 15.3 | 95.5 |
| Natural sciences | 83.6 | 86.3 | 18.0 | 93.6 |
| All sciences | 78.3 | 87.7 | 18.1 | 94.4 |

SOURCE: Data are from the National Research Council, Office of Scientific and Engineering Personnel's biennial surveys of doctorate recipients.

- Although the differences may not be statistically significant, it appears that relative to other areas of science a slightly higher percentage of agricultural scientists are more than 55 years old and a slightly lower percentage are U.S. citizens (Table 3-12). These figures may indicate a declining interest among U.S. students in pursuing agricultural science careers, while the age figure suggests that new leadership opportunities are at hand. The higher percent of non-U.S. citizens in agricultural sciences may also suggest that the colleges of agriculture have a relatively larger role in international training than do nonagriculture schools.

- Table 3-13 presents trends in the numbers of women and ethnic minorities in the agricultural sciences. Clearly, women are a significantly larger percentage in agricultural science today than 20 years ago; however, their presence is still minimal in relation to that of women in either life or natural sciences.
- In contrast, ethnic minorities are represented in similar proportions in agricultural sciences and other areas of science. In all areas of science, participation by ethnic minorities has grown more slowly than participation of women. It is interesting that only in the agricultural sciences are ethnic minorities actually better represented than women as a group (Table 3-13). (This, however, may reflect inclusion of U.S.-educated scientists from, for example, Africa, Latin America, Asia, etc.)
- Table 3-14 shows that the percent of agricultural scientists less than 35 years old is declining as it is in all areas of science. Thus the science community is aging generally, although agricultural scientists are somewhat older than their peers.

TABLE 3-13 Women and Ethnic Minority Scientists (percent) Holding Doctorate Degrees, 1973, 1979, 1985, 1991

| | Year | | | |
|---|---|---|---|---|
| Discipline | 1973 | 1979 | 1985 | 1991 |
| *Women* | | | | |
| Agricultural sciences | 1.3 | 2.6 | 5.5 | 9.3 |
| Agricultural economics | 0.6 | 1.7 | 5.2 | 12.5 |
| Animal sciences | 1.1 | 2.7 | 4.8 | 8.4 |
| Plant and soil sciences | 1.1 | 1.8 | 4.9 | 7.9 |
| Forestry and wildlife management | 0.0 | 0.2 | 2.2 | 7.1 |
| Food sciences | 7.0 | 16.3 | 17.8 | 18.4 |
| Life sciences | 12.7 | 15.8 | 21.3 | 26.7 |
| Natural sciences | 6.8 | 8.9 | 12.5 | 16.4 |
| All sciences | 8.9 | 12.3 | 17.0 | 21.6 |
| *Ethnic Minorities* | | | | |
| Agricultural sciences | 6.3 | 11.0 | 12.0 | 11.8 |
| Agricultural economics | 7.0 | 13.3 | 15.1 | 11.5 |
| Animal sciences | 2.3 | 6.5 | 8.2 | 9.4 |
| Plant and soil sciences | 7.7 | 12.2 | 11.5 | 10.1 |
| Forestry and wildlife management | 1.1 | 4.1 | 5.1 | 5.9 |
| Food sciences | 12.7 | 30.2 | 32.8 | 35.4 |
| Life sciences | 6.3 | 8.8 | 9.8 | 11.3 |
| Natural sciences | 5.7 | 9.0 | 10.8 | 12.4 |
| All sciences | 5.4 | 8.1 | 9.9 | 11.2 |

SOURCE: Data are from the National Research Council, Office of Scientific and Engineering Personnel's biennial surveys of doctorate recipients.

TABLE 3-14 Scientists (percent) Less Than 35 Years Old Holding Doctorate Degrees, 1973, 1979, 1985, 1991

| | Year | | | |
|---|---|---|---|---|
| Discipline | 1973 | 1979 | 1985 | 1991 |
| Agricultural sciences | 21.4 | 13.7 | 13.2 | 8.8 |
| Agricultural economics | 56.3 | 24.7 | 20.1 | 11.1 |
| Animal sciences | 17.6 | 12.1 | 13.8 | 10.8 |
| Plant and soil sciences | 20.5 | 11.8 | 13.2 | 8.2 |
| Forestry and wildlife management | 20.3 | 13.9 | 7.3 | 4.4 |
| Food sciences | 26.8 | 22.6 | 13.5 | 11.9 |
| Life sciences | 27.1 | 22.6 | 16.5 | 10.6 |
| Natural sciences | 28.3 | 20.4 | 15.2 | 11.5 |
| All other sciences | 27.3 | 20.6 | 14.4 | 10.1 |

SOURCE: Data are from the National Research Council, Office of Scientific and Engineering Personnel's biennial surveys of doctorate recipients.

## ISSUES FOR DISCUSSION

- What factors explain the trends in enrollment at colleges of agriculture? Will the large agribusiness sector, and strong public interest in food, natural resource, and environmental issues, generate a continued demand for programs of colleges of agriculture even though few students will enter or return to farming? Are college curricula adjusting to meet the interests of today's students and the needs of today's agribusiness industry?
- Do the instruction programs of the colleges provide the basic knowledge and practical skills pertinent to those who do return to farms or join businesses that serve farms? (Are they too "discipline-oriented"?)
- Are programs of the colleges adjusting or consolidating in accordance with trends in student demand? Could system-wide efficiencies be realized through increased specialization by individual colleges in the offering of specific degrees?
- What are the future relative roles of land grant colleges of agriculture and non-land grants in educating students in agriculture, food, and natural resource fields?
- Ph.D. students have different academic program emphases than undergraduates. Is the pool of potential new faculty appropriate to the future instructional needs of the colleges?
- What qualities and characteristics are desired of the current and future leadership of the colleges of agriculture?
- Given the growing diversity of the system's clientele, how important to the system's future is a diversity of backgrounds and views among its leadership? How can the colleges enhance opportunities for and attractiveness of their programs to women, minorities, and young scientists?

## SUGGESTED READINGS

Coulter, K. Jane, Allan D. Goecker, and Marge Stanton. *Employment Opportunities for College Graduates in the Food and Agricultural Sciences: Agriculture, Natural Resources, and Veterinary Medicine.* Washington, D.C.: Higher Education Programs, Cooperative State Research Service, U.S. Department of Agriculture , 1990.

Marchant, Mary, and Handy Williamson, Jr., eds. *Achieving Diversity: The Status and Progress of Women and African Americans in the Agricultural Economics Profession.* New York: Garland Publishing, Inc., 1994.

National Research Council. *Agriculture and the Undergraduate.* Washington, D.C.: National Academy Press, 1992.

National Research Council. *Doctorate Recipients from United States Universities: Summary Report 1992.* Washington, D.C.: National Academy Press, 1993.

National Research Council. *Educating the Next Generation of Agricultural Scientists.* Washington, D.C.: National Academy Press, 1988.

National Research Council. *Plant Biology Research and Training for the 21st Century.* Washington, D.C.: National Academy Press, 1992.

National Research Council. *Understanding Agriculture: New Directions for Education.* Washington, D.C.: National Academy Press, 1988.

National Research Council. *Women Scientists and Engineers Employed in Industry: Why So Few*? Washington, D.C.: National Academy Press, 1994.

U.S. Department of Education, National Center for Education Statistics. *The Condition of Education 1993.* Washington, D.C.: U.S. Department of Education, 1993.

# 4

# RESEARCH AT LAND GRANT COLLEGES OF AGRICULTURE: The State Arm of the U.S. Public Agricultural Research System

*This chapter introduces the second of the land grant colleges' functions, that of agricultural research. The colleges are the state-based component of the public agricultural research system; the federal component includes the intramural science agencies of the U.S. Department of Agriculture (USDA). The chapter compares USDA funding for intramural and extramural research and compares USDA funding for university-based agricultural research to funding for university research by other federal agencies. The research is described generally, in terms of level of expenditures and area of emphasis. The description of research expenditures is drawn from the USDA's Current Research Information System (CRIS), an inventory of agricultural research based on reports filed by research scientists and administrators.*

- The public agricultural research system may be unique among U.S. science institutions. Responsibility for publicly funded agricultural research is divided between scientists employed by USDA and those employed by the state agricultural experiment stations and other units of universities. Agencies of USDA both conduct research—in fact they receive the lion's share of public agricultural research funds —and administer research and extension grants.
- USDA's Agricultural Research Service (ARS), the National Agricultural Library (NAL), the research units of the Forest Service (FS), and the Economic Research Service (ERS) compose the intramural research agencies. The Cooperative State Research, Education, and Extension Service (CSREES) administers the partnership grant programs that fund extramural research, cooperative extension, instructional support, and training.[1]

[1] The recent USDA reorganization groups research, education, extension, and economics agencies under an undersecretary for research, education, and economics. The Cooperative State Research Service (CSRS) and the Extension Service were merged to form the Cooperative State Research, Education, and Extension Service (CSREES). The ARS and ERS remain separate agencies under this undersecretary. The Forest Service, along with the Natural Resources Conservation Service, reports to the undersecretary for natural resources and environment.

- As Tables 4-1 and 4-2 indicate, in 1993 30 percent of USDA's research appropriations supported extramural research—that is, research supported by but not conducted by USDA, most of which occurs at the state agricultural experiment stations (SAESs) associated with the 1862 land grant colleges of agriculture or at forestry and veterinary medicine schools. This compares with 27 percent in 1980 and 29 percent in 1990. It would seem that the importance of extramural research, in terms of all federally funded agricultural research, is increasing, if only slightly. As noted in Chapter 6, an expanding source of federal funds for extramural agricultural research has been the competitive grants program known as the National Research Initiative (NRI). The NRI, however, is open to scientists both inside and outside the SAES system; thus the full amount of this increase has not gone to USDA's traditional extramural partners at the land grant colleges of agriculture.
- A comparison of ARS and CSRS appropriations shows that between 1980 and 1995 the budget for intramural research grew at a slightly slower pace than that for extramural research. CSRS appropriations grew 108 percent, an annual average rate of 7.2 percent in nominal dollars. ARS appropriations grew 98 percent, or 6.5 percent per year (Table 4-1). However, in both cases, *real* budget increases—that is, accounting for the effects of inflation—were much smaller. The real value of total USDA research agency appropriations increased less than 1 percent annually (average) between 1980 and 1990 and only 2 percent annually (average) between 1990 and 1993 (Table 4-2).

TABLE 4-1 Federal Appropriations (current dollars in millions) for USDA Research Agencies, 1980–1995

| | Intramural | | | | | |
|---|---|---|---|---|---|---|
| Year | ARS | FS | ERS | NAL | CSRS[a] | Total |
| 1980 | 358.0 | 95.9 | 35.2 | 7.3 | 185.9 | 682.3 |
| 1981 | 404.1 | 108.4 | 39.5 | 8.2 | 200.7 | 760.9 |
| 1982 | 423.2 | 112.1 | 39.4 | 8.2 | 220.6 | 803.5 |
| 1983 | 451.9 | 107.7 | 38.8 | 9.1 | 232.3 | 839.8 |
| 1984 | 471.1 | 109.4 | 44.3 | 10.4 | 237.7 | 872.9 |
| 1985 | 491.4 | 113.8 | 46.6 | 11.5 | 284.4 | 947.7 |
| 1986 | 483.2 | 113.6 | 44.1 | 10.8 | 269.6 | 921.3 |
| 1987 | 511.4 | 126.7 | 44.9 | 11.1 | 293.7 | 987.8 |
| 1988 | 544.1 | 132.5 | 48.3 | 12.2 | 303.1 | 1,040.2 |
| 1989 | 569.4 | 138.3 | 49.6 | 14.3 | 290.8 | 1,062.4 |
| 1990 | 593.3 | 150.9 | 51.0 | 14.7 | 326.6 | 1,136.5 |
| 1991 | 631.0 | 167.6 | 54.4 | 16.8 | 373.3 | 1,243.1 |
| 1992 | 668.4 | 180.5 | 58.7 | 17.8 | 415.5 | 1,340.9 |
| 1993 | 668.0 | 182.1 | 58.9 | 17.7 | 401.7 | 1,328.5 |
| 1994 | 691.6 | 192.5 | 55.3 | 18.3 | 423.1 | 1,380.7 |
| 1995[b] | 708.6 | 204.0 | 53.7 | 19.7 | 386.9 | 1,372.9 |

Abbreviations: ARS, Agricultural Research Service; FS, Forest Service; ERS, Economic Research Service; NAL, National Agricultural Library; CSRS, Cooperative State Research Service.

[a] CSRS appropriations are for extramural research.

[b] President's budget request.

SOURCE: Adapted from National Research Council. 1994. Investing in the National Research Initiative. Washington, D.C.: National Academy Press.

TABLE 4-2 Federal Appropriations (real dollars in millions) for USDA Research Agencies, 1980–1993 (1987 = 100)

| | Intramural | | | | | |
|---|---|---|---|---|---|---|
| Year | ARS | FS | ERS | NAL | CSRS[a] | Total |
| 1980 | 497.2 | 133.2 | 48.9 | 10.1 | 258.2 | 947.6 |
| 1981 | 513.5 | 137.7 | 50.2 | 10.4 | 255.0 | 966.8 |
| 1982 | 503.8 | 133.5 | 46.9 | 9.8 | 262.6 | 956.5 |
| 1983 | 515.3 | 122.8 | 44.2 | 10.4 | 264.9 | 957.6 |
| 1984 | 515.4 | 119.7 | 48.5 | 11.4 | 260.1 | 955.0 |
| 1985 | 517.3 | 119.8 | 49.1 | 12.1 | 299.4 | 997.6 |
| 1986 | 496.1 | 116.6 | 45.3 | 11.1 | 276.8 | 945.9 |
| 1987 | 511.4 | 126.7 | 44.9 | 11.1 | 293.7 | 987.8 |
| 1988 | 525.2 | 127.9 | 46.6 | 11.8 | 292.6 | 1,004.1 |
| 1989 | 528.2 | 128.3 | 46.0 | 13.3 | 269.8 | 985.5 |
| 1990 | 528.3 | 134.4 | 45.4 | 13.1 | 290.8 | 1,012.0 |
| 1991 | 543.0 | 144.2 | 46.8 | 14.5 | 321.3 | 1,069.8 |
| 1992 | 558.4 | 150.8 | 49.0 | 14.9 | 347.1 | 1,120.2 |
| 1993 | 541.8 | 147.7 | 47.8 | 14.4 | 325.8 | 1,077.5 |

NOTE: The deflating index used is the implicit price deflator for total government purchases of goods and services. Abbreviations: ARS, Agricultural Research Service; FS, Forest Service; ERS, Economic Research Service; NAL, National Agricultural Library; CSRS, Cooperative State Research Service.

[a] CSRS appropriations are for extramural research.

Colleges of agriculture are the state-based component of a federal-state partnership in agricultural research.

- The SAESs at land grant colleges of agriculture have traditionally been the state-based partners in the public agricultural research system in the United States. As such, they have been the principal recipients of USDA appropriations for extramural research. Table 4-3 shows that in 1981 USDA appropriations earmarked for research at universities and colleges were 5.5 percent of all such federal agency obligations. By 1991 these USDA obligations had fallen to 4.7 percent.
- Agricultural research at universities accounts for only a small percentage of all federally funded university research; however, that share is still larger relative to the percent of all students enrolled at colleges of agriculture. This may be an indication that colleges of agriculture are relatively research intensive in relation to other university colleges and departments.
- To expand or even maintain their research programs, agriculture colleges have had to seek out other funding sources. That some have done so, reducing over time USDA's traditional role as the federal partner in agricultural research, is the subject of Chapter 6.

TABLE 4-3 Federal Agency Appropriations (millions of dollars) for Research and Development at Universities and Colleges, 1966–1991

| Agency | 1966 | 1971 | 1976 | 1981 | 1986 | 1991 |
|---|---|---|---|---|---|---|
| Agency for International Development | 0 | 17 | 11 | 31 | 44 | 32 |
| Department of Agriculture | 63 | 75 | 124 | 241 | 290 | 409 |
| Department of Commerce | 2 | 7 | 29 | 51 | 68 | 62 |
| Department of Defense | 278 | 249 | 212 | 700 | 1,055 | 0 |
| Department of Education | 0 | 0 | 0 | 56 | 72 | 92 |
| Department of Energy | 83 | 96 | 138 | 282 | 347 | 605 |
| Department of Health and Human Services | 507 | 696 | 1,296 | 2,113 | 3,212 | 5,301 |
| Department of Housing and Urban Development | 0 | 1 | 1 | 3 | 0 | 0 |
| Department of the Interior | 19 | 21 | 26 | 30 | 41 | 68 |
| Department of Labor | 0 | 3 | 3 | 13 | 4 | 6 |
| Department of Transportation | 0 | 8 | 15 | 28 | 13 | 35 |
| Environmental Protection Agency | 0 | 17 | 29 | 64 | 68 | 104 |
| National Aeronautics and Space Administration | 107 | 129 | 107 | 174 | 254 | 534 |
| National Science Foundation | 192 | 217 | 437 | 617 | 984 | 1,439 |
| Nuclear Regulatory Commission | 0 | 0 | 3 | 7 | 4 | 5 |
| Office of Economic Opportunity | 0 | 16 | 0 | 0 | 0 | 0 |
| Total | 1,252 | 1,552 | 2,431 | 4,411 | 6,456 | 8,691 |

SOURCE: Data provided by the National Science Foundation.

## COLLEGE EXPENDITURES FOR RESEARCH

- Each year SAESs and other institutions receiving funds from grant programs administered by USDA must report their total research expenditures, including those based on non-USDA sources of support, to USDA. The reports, which are based on research project descriptions filed by individual research scientists, are entered into a computerized data base known as the Current Research Information System (CRIS). CRIS is maintained and updated by CSREES (formerly CSRS) and can be used to analyze trends in agricultural research expenditures, including the size of expenditures over time and the distribution of expenditures among areas of research investigation, academic discipline, and individual colleges (or experiment stations). CRIS can also be used to analyze sources of support for agricultural research, which is the topic of Chapter 6.
- CRIS is most thorough for SAESs at 1862 and 1890 colleges of agriculture because institutions in these classifications must report research activities as expenditures to USDA. However, CRIS also includes data on research expenditures by a significant number of the forestry schools and veterinary colleges. Forestry schools that receive funds through the McIntire-Stennis Forestry Research Act of 1962, for example, must report their total research expenditures to USDA, which incorporates the data in CRIS.
- CRIS data cover approximately 20 years, beginning with the early 1970s. Table 4-4 shows that reported research expenditures have increased (without adjusting for inflation) fivefold for 1862s, approximately threefold for 1890s, and more than tenfold for forestry schools since the early 1970s. Total expenditures did not quite double between 1982 and 1992; thus research activities increased more rapidly (at least in nominal terms) prior to the 1980s. The numbers in Table 4-4 indicate clearly that 1862 colleges account for the vast majority of research in agriculture, forestry, and animal health.

TABLE 4-4 Total Research Expenditures (millions of dollars), by Institution Classification, 1972–1992

| | Year | | | | |
|---|---|---|---|---|---|
| Institution Classification | 1972 | 1977 | 1982 | 1987 | 1992 |
| 1862 SAESs | $359 | $606 | $973 | $1,312 | $1,772 |
| 1890 Colleges and Universities | 9 | 13 | 19 | 21 | 30 |
| Forestry Schools | 6 | 17 | 25 | 32 | 64 |
| Schools of Veterinary Medicine | NA | NA | 52 | 82 | 125 |
| Total | $374 | $636 | $1,067 | $1,448 | $1,992 |

NOTE: Number and combination of reporting institutions may vary from year to year; comparison of data across years should be made with caution. NA, data not available because institution class did not report.
SOURCE: Data are from USDA Current Research Information System (CRIS).

What types of research do scientists at colleges of agriculture do?

- The Experiment Station Committee on Organization and Policy (ESCOP), the system-wide research planning committee, designates six main areas of agricultural research:

1. environment and natural resources;
2. nutrition, food safety, and health;
3. processes and products;
4. economic and social issues;
5. animal systems; and
6. plant systems.

One way that research projects reported to CRIS are categorized is according to these six ESCOP program areas (see box copy, p. 63).

- Approximately 62 percent of the system's research expenditures are allocated to the plant ($\approx$ 35 percent) and animal ($\approx$ 27 percent) systems research program areas (Table 4-5). The broad goals of the majority of the research in these program areas are to better protect crops and livestock from insects, disease, and other hazards and to maintain adequate food production at decreasing real production costs (U.S. Department of Agriculture, 1993a). Thus a majority of the colleges' research is aimed at improving the output and productivity of crop and animal production on farms.
- Another 18 percent of the system's research expenditures are allocated to research projects in the environment and natural resources category. A significant portion of this research is aimed at improving management of forest resources and protecting forests from insects, disease, and other hazards. Thus much of this research has a strong parallel to that conducted in the animal and plant systems areas (U.S. Department of Agriculture, 1993a).
- The benefits of animal and plant systems research can accrue to consumers through lower food prices and healthier food; for example, animal health research can result in lower risks to humans from animal-transmitted disease. However, CRIS data indicate that only 3 percent of the system's research directly addresses consumer health and safety through the study of nutrition or food safety and quality. A closer examination of specific research efforts is needed to assess how plant and animal systems research affects the nutritional quality or safety of food.

**ESCOP Research Program Areas**

The strategic planning process for the state agricultural experiment stations (SAESs) is conducted by the Experiment Station Committee on Organization and Policy (ESCOP)—the executive body of the Experiment Station Section of the National Association of State Universities and Land Grant Colleges. The ESCOP Strategic Planning Subcommittee is responsible for preparing the strategic agenda for the SAES system, with annual updates and a major revision every 4 years.

The subcommittee solicits input from producers, consumer groups, and the science community through conferences and individual surveys. The process also brings the state and federal agencies into a "coordinated frame of reference" in terms of communicating the research priorities for U.S. agricultural science to key policy and decision makers. The research initiatives for 1994 are listed here, grouped into six categories:

1. Environment and Natural Resources
   - Conserve and enhance air, soil, and water resources
   - Manage ecosystems to conserve and enhance biodiversity
   - Recover and use waste resources through agricultural and forestry systems
   - Develop resource management decision systems
2. Nutrition, Food Safety, and Health
   - Enhance food safety
   - Target optimal nutrition for individual health
   - Design foods for healthy diets
   - Promote healthy food choices
3. Processes and Products
   - Convert processing byproducts to beneficial uses
   - Enhance food quality and value
   - Develop new or improved non-food products
4. Economic and Social Issues
   - Enhance agricultural and rural economies
   - Strengthen communities
   - Empower people for economic and social viability
5. Animal Systems
   - Develop integrated/sustainable animal production systems
   - Enhance animal genetic diversity and biological performance
   - increase the quality of animal food products
   - Enhance the health and well-being of food animals
6. Plant Systems
   - Protect plants for sustained productivity
   - Develop alternative plant management systems
   - Understand fundamental plant processes
   - Use genetics to improve plants for the 21st century

SOURCE: Experiment Station Committee on Organization and Policy. 1994. Opportunities to Meet Changing Needs: Research on Food, Agriculture, and Natural Resources. Washington, D.C.: U.S. Department of Agriculture.

TABLE 4-5 Research Expenditures (thousands of dollars) by Institution Classification and ESCOP Program Area, 1992

| | ESCOP Program Area | | | |
|---|---|---|---|---|
| Institution Classification | Environment and Natural Resources | Plant Systems | Animal Systems | Food and Nutrition for Optimal Health |
| 1862 SAESs | $307,006 (17.32) | $695,748 (39.25) | $404,751 (22.84) | $63,084 (3.56) |
| 1890 Colleges and Universities | 3,502 (11.69) | 8,649 (28.86) | 8,121 (27.10) | 2,962 (9.88) |
| Forestry Schools | 45,818 (71.21) | 699 (1.09) | 25 (0.04) | 0 (0.00) |
| Schools of Veterinary Medicine | 3,847 (3.08) | 262 (0.21) | 118,119 (94.52) | 157 (0.13) |
| Total | $360,173 (18.08) | $705,358 (35.41) | $531,017 (26.66) | $66,203 (3.32) |

NOTE: Number in parentheses is percent of the total for institution class.
SOURCE: Data are from USDA Current Research Information System (CRIS). (*table continued on next page*)

- Similarly, only 10 percent of the system's research directly addresses social science concerns such as farm and rural income and community development. Plant, animal, and forestry research can, of course, affect rural economies and communities through its effects on farm and forest productivity and farm sector structure, but ESCOP classifications do not make these linkages explicit.
- Figure 4-1 shows that there has been little shift in the allocation of research expenditures since 1972. A more precise assessment of specific research programs and projects would be required, however, to determine whether research is in fact shifting toward areas that currently concern consumer, environmental, sustainable agriculture, and rural development groups. Decreasing the use of chemicals in agricultural production, enhancing sustainable agriculture systems, reconciling diet and human health, and enhancing small-farm and rural community viability typify some concerns.
- At the 1890 colleges a larger percentage of research is devoted to food, nutrition, and social science issues—29 percent at 1890s versus 14 percent at 1862s. It may be that because of their history and location—often in primarily rural, poor African American communities—1890s view their service mission as one that is more specifically aimed at the concerns of rural households, limited-resource farmers, and economically disadvantaged populations (Table 4-5).

TABLE 4-6 Scientist Years (full-time equivalents) by Institution Classification and ESCOP Program Area, 1992

| | ESCOP Program Area | | | |
|---|---|---|---|---|
| Institution Classification | Environment and Natural Resources | Plant Systems | Animal Systems | Food and Nutrition for Optimal Health |
| 1862 SAESs | 1,107 (17.21) | 2,597 (40.38) | 1,229 (19.11) | 229 (3.56) |
| 1890 Colleges and Universities | 23 (13.53) | 54 (31.76) | 33 (19.41) | 18 (10.59) |
| Forestry Schools | 198 (71.74) | 1 (0.36) | 0 | 0 |
| Schools of Veterinary Medicine | 13 (2.80) | 1 (0.22) | 439 (94.41) | 2 (0.43) |
| Total | 1,340 (18.25) | 2,654 (36.15) | 1,701 (23.17) | 249 (3.39) |

NOTE: Number in parentheses is percent of the total for institution class.
SOURCE: Data are from USDA Current Research Information System (CRIS).

| | ESCOP Program Area | | | |
|---|---|---|---|---|
| Institution Classification | Processing for Added Value | Social Sciences Issues | Unassigned | Total Research Expenditures |
| 1862 SAESs | $93,822 (5.29) | $177,633 (10.02) | $30,421 (1.72) | $1,772,467 |
| 1890 Colleges and Universities | 968 (3.23) | 5,562 (18.56) | 203 (0.68) | 29,966 |
| Forestry Schools | 11,221 (17.44) | 5,120 (7.96) | 1,454 (2.26) | 64,338 |
| Schools of Veterinary Medicine | 34 (0.03) | 2,123 (1.70) | 420 (0.34) | 124,961 |
| Total | $106,046 (5.32) | $190,439 (9.56) | $32,498 (1.63) | $1,991,733 |

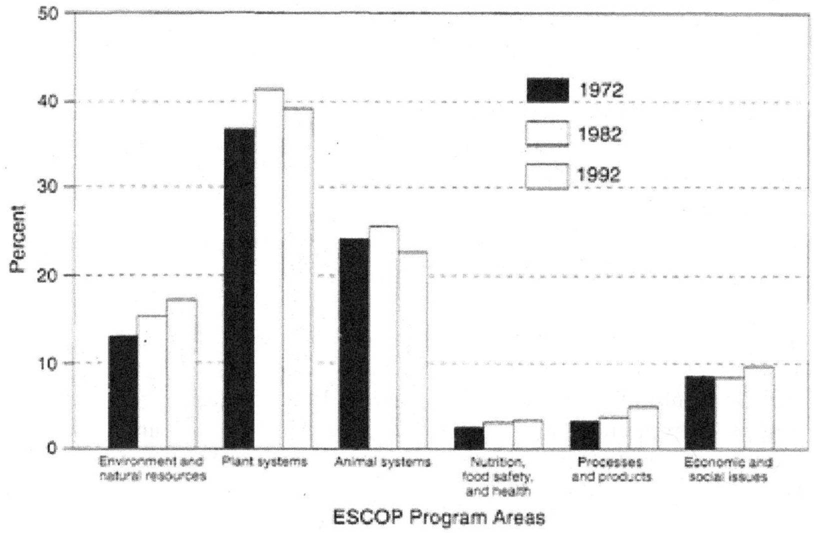


**FIGURE 4-1** *Since 1972 there has been little change in the relative distribution (percent) of research expenditures to ESCOP program areas at 1862 institutions. Between 1982 and 1992 there were small shifts away from animal and plant sciences and toward research in natural resources, social sciences, processing and value added, and food and nutrition areas.*

| | ESCOP Program Area | | | |
|---|---|---|---|---|
| Institution Classification | Processing for Added Value | Social Sciences Issues | Unassigned | Total Scientist Years |
| 1862 SAESs | 332 (5.16) | 826 (12.84) | 112 (1.74) | 6,432 |
| 1890 Colleges and Universities | 8 (4.17) | 33 (19.41) | 1 (0.59) | 170 |
| Forestry Schools | 48 (17.39) | 26 (9.42) | 4 (1.45) | 276 |
| Schools of Veterinary Medicine | 0 | 9 (1.94) | 1 (0.22) | 465 |
| Total | 388 (5.28) | 893 (12.16) | 118 (1.61) | 7,342 |

- Like research monies, the time of research scientists at colleges of agriculture is allocated predominately toward research initiatives in plant, animal, and natural resource systems. In 1992, 229 of the 6,432 "scientist years"[2] reported by SAESs at 1862 colleges were devoted to studying nutrition, health, or food safety (Table 4-6).
- Not surprising is the fact that distribution of research scientists' time corresponds to the distribution of doctorate students in the various specializations noted in Chapter 3. In 1992, 1,002 of 1,860 doctorate students (54 percent) obtained degrees in animal and plant sciences and "related" sciences (see Table 3-6). In that same year, 59 percent of research scientist years were devoted to investigation of plant and animal systems (Table 4-6). Note also that in 1992 only 6,145 of 19,426 bachelor's degrees (32 percent) were awarded in fields of plant, animal, and related sciences (see Table 3-6).

## ISSUES FOR DISCUSSION

- The colleges of agriculture compose one component of the U.S. public agricultural research system, which also includes agencies of USDA. How does the colleges' role differ from or complement that of the federal agencies? Should extramural (university-based) research be a larger component of the system than it now is?
- USDA appropriations for agricultural research now represent a small (and declining) fraction of total federal appropriations for research and development at U.S. colleges and universities (though economic studies show very high rates of return to public investments in agricultural research). What criteria could be used to determine whether this share is commensurate with the importance of the agricultural complex to the U.S. economy and social welfare?
- Since 1975 research information data show relatively little change in the allocation of college research dollars to research program areas. Should more change have occurred, and does the agricultural research inventory system that is currently used mask real change in research emphases or goals that may have occurred?

## SUGGESTED READINGS

Alston, J. M., G. W. Norton, and P. G. Pardey. *Science Under Scarcity: Principles and Practice for Agricultural Research Evaluation and Priority Setting*. Ithaca, NY: Cornell University Press, 1995.

Congress of the United States, Office of Technology Assessment. *A New Technological Era for American Agriculture*. Washington, D.C.: U.S. Government Printing Office, 1992.

Congress of the United States, Office of Technology Assessment. *An Assessment of the United States Food and Agricultural Research System* . Washington, D.C.: U.S. Government Printing Office, 1981.

Congress of the United States, Office of Technology Assessment. *Federally Funded Research: Decisions for a Decade*. Washington, D.C.: U.S. Government Printing Office, May 1991.

Huffman, Wallace E., and Robert E. Evenson. *Science for Agriculture: A Long-Term Perspective*. Ames: Iowa State University Press, 1993.

Miller, Raymond J., and Clare I. Harris. *Trends in Agricultural Research: Thoughts for Discussion*, Cooperative State Research Service. Washington, D.C.: U.S. Department of Agriculture, 1994.

U.S. Department of Agriculture, Economic Research Service. *Chronological Landmarks in American Agriculture*, AIB No. 425. Washington, D.C.: U.S. Department of Agriculture, 1990.

---

[2] "Scientist years" represent the number of years of effort (rounded to the nearest whole year) devoted by researchers in the SAESs and other cooperating institutions.

# 5

# THE EVOLUTION OF EXTENSION AT THE LAND GRANT COLLEGES OF AGRICULTURE

*This chapter introduces the third aspect of the colleges' tripartite mission, that of off-campus extension. It describes the funding base for extension, the geographic allocation of extension resources, and the allocation of extension resources among major program emphases. The data, provided by the U.S. Department of Agriculture's Extension Service (now a component of the Cooperative State Research, Education, and Extension Service), is used to compare the types of problems being addressed by extension staff in relation to those of interest to the colleges' research scientists.*

- The theory behind university extension is that education and research developments achieved through public funding should be more broadly available to those not attending the institutions and throughout one's lifetime. To realize that goal, programs were developed that geographically extended the availability of the educational resources of an institution by special arrangements such as correspondence courses and on-site consultations to persons otherwise unable to take advantage of such resources. The concept of "university extension" was introduced by U.S. colleges and universities working through city libraries. In the 1890s New York appropriated funds for university extension work and the University of Chicago included extension in its original plan of organization.
- Agricultural colleges also began to look at the extension movement in the 1890s. For example, Rutgers-The State University of New Jersey offered six lectures each on soils and crops, feeding plants, and animal nutrition at different locations around the state.
- The 1914 Smith Lever Act established extension on a nationwide basis as a unique cooperative effort by federal, state, and local governments. The federal mandate came in response to concerns that information and technology being developed at the SAESs and USDA were not reaching many farmers, particularly those most in need of education. The colleges and SAESs were understaffed in relation to needs, and a gap was developing between professors on the campus and farmers in the fields.

- In the decades that followed, this third function of the land grant colleges may have faced greater pressures for change than either campus instruction or research; and because of its relatively strong local base of support, it may have reoriented itself in response to local needs more than either research or campus-based teaching programs. The extension program's original mandate—to educate farmers regarding new farm technologies and ways in which farm life could be improved—has been challenged by a number of factors, principally the decline in farm population and the changing profile of farms and farmers themselves.
- In 1988 the Cooperative Extension Service reformulated its statement of purpose to stress its role in helping people help themselves "through an educational process which uses scientific knowledge focused on issues and needs" (Rasmussen, 1989: p. 223). The statement is unbounded by discipline, audience, or geography, leaving open the question of who, primarily, extension should serve. In today's context, some have asked whether state and local extension services should continue to draw primarily from the research and programs of the colleges of agriculture or, instead, become a conduit for the research and programs of the entire university.

Although the role of the federal partnership has been declining, federal funds are increasingly earmarked for specific extension activities.

- Table 5-1 shows that total cooperative extension funding to states grew more slowly during the 1982 to 1992 period than during the previous decade. At the same time, the role of the federal partner in providing extension services has been declining. Twenty years ago federal funds accounted for 42 percent of all funding; in 1992 the federal funds were only 29 percent (Figure 5-1). In fact, these numbers may understate the declining role of the federal and even the public sector; there is evidence of a growing role for private-sector firms in providing extension-type services to farmers in particular (Bradshaw and Marquart, 1990) (see box copy, p. 70).

TABLE 5-1 Sources of Funds (millions of dollars) Allocated to States for Cooperative Extension Work, 1972–1992

| | Year | | | | |
|---|---|---|---|---|---|
| Source | 1972 | 1977 | 1982 | 1987 | 1992 |
| Federal | $149 | $198 | $302 | $319 | $401 |
| State | 136 | 220 | 368 | 500 | 652 |
| Local | 70 | 105 | 182 | 229 | 333 |
| Total | $354 | $524 | $852 | $1,048 | $1,386 |

SOURCE: Data were provided by the USDA Extension Service.

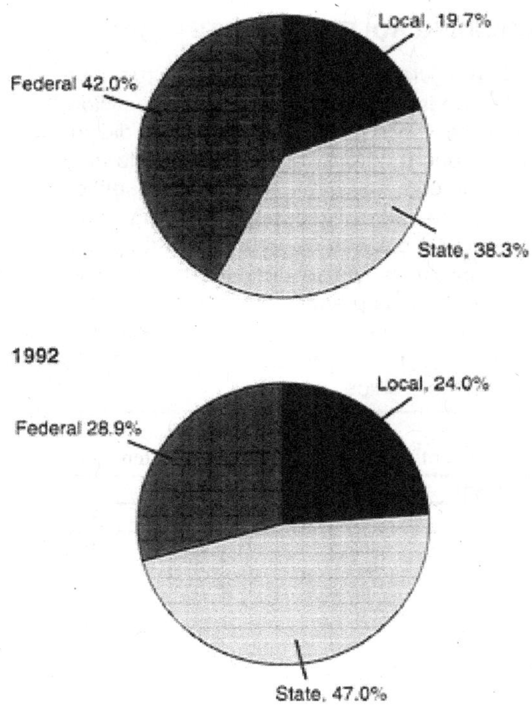


**FIGURE 5-1**
*Between 1972 and 1992, federal funding of cooperative extension services decreased from 42 percent to 29 percent of total funding.*

- As the role for the federal partner has declined, Congress' role in directing extension programs has increased. Table 5-2 shows that in the last 5 years formula funds for extension have grown more slowly than "special" funds. Special funds are those earmarked by Congress for specific types of services. Urban and rural nutrition programs, such as the Expanded Food and Nutrition Education Program (EFNEP), have been a particular focus of Congressional earmarks. Water quality, pest management, and youth at risk have also been targets of earmarked funds. On the other hand, federal support for rural and community development programs has been inconsistent (Rasmussen, 1989).

Extension staff divide their time among farm service, community development, and consumer education programs, while research scientists target crop and animal production.

## THE EXPANDING ROLE OF PRIVATE CROP CONSULTANTS

According to the American Association of Independent Crop Consultants, independent consultants now work with farmers on production issues, such as nutrient, pesticide, and fungicide requirements, on a one-to-one basis in much the same way that extension agents did in earlier years. These independent agents draw on the resources provided by the extension specialists at colleges of agriculture. They look to the Cooperative Extension Service to coordinate many of the farm service functions in their geographic area.

Independent crop consultants are typically educated at the land grant colleges because they are required to obtain a 4-year agricultural science degree before they can be certified. The association has produced position papers promoting revisions to curricula at the colleges, revisions designed to make coursework more pertinent to the in-the-field practice of crop consulting (Bradshaw and Marquart, 1990).

The independent crop consultant business really got off the ground during the 1970s, although private consultants in cotton-producing regions have been active for 40 years. Other private crop consultants work for the fertilizer and other chemical dealerships.

TABLE 5-2 USDA Appropriations (millions of dollars) for Cooperative Extension

| | Funding Mechanism | | | |
|---|---|---|---|---|
| Year | Formula | Special[a] | Other | Total |
| 1980 | 200.7 | 78.3 | 6.5 | 285.5 |
| 1981 | 217.6 | 80.1 | 5.9 | 303.6 |
| 1982 | 232.6 | 76.8 | 6.3 | 315.7 |
| 1983 | 247.6 | 75.6 | 5.4 | 328.6 |
| 1984 | 253.2 | 75.6 | 5.5 | 334.3 |
| 1985 | 260.2 | 77.6 | 5.9 | 343.7 |
| 1986 | 260.2 | 78.9 | 5.5 | 344.6 |
| 1987 | 254.1 | 78.6 | 6.3 | 339.0 |
| 1988 | 260.8 | 80.2 | 16.9 | 357.9 |
| 1989 | 260.8 | 82.0 | 18.6 | 361.4 |
| 1990 | 265.1 | 86.4 | 18.2 | 369.7 |
| 1991 | 276.4 | 103.4 | 18.7 | 398.5 |
| 1992 | 288.5 | 110.0 | 20.9 | 419.4 |
| 1993 | 288.5 | 118.0 | 18.4 | 424.9 |
| 1994 | 298.1 | 117.4 | 19.1 | 434.6 |
| 1995[b] | 298.1 | 121.4 | 13.2 | 432.7 |

NOTE: Totals include appropriations for other federal agencies, federal administration, legislative set-asides, and allocations to states.

[a] "Special" funds are those earmarked by Congress for specific types of services.

[b] President's budget request.

SOURCE: Data were provided by the USDA Office of Budget and Program Analysis.

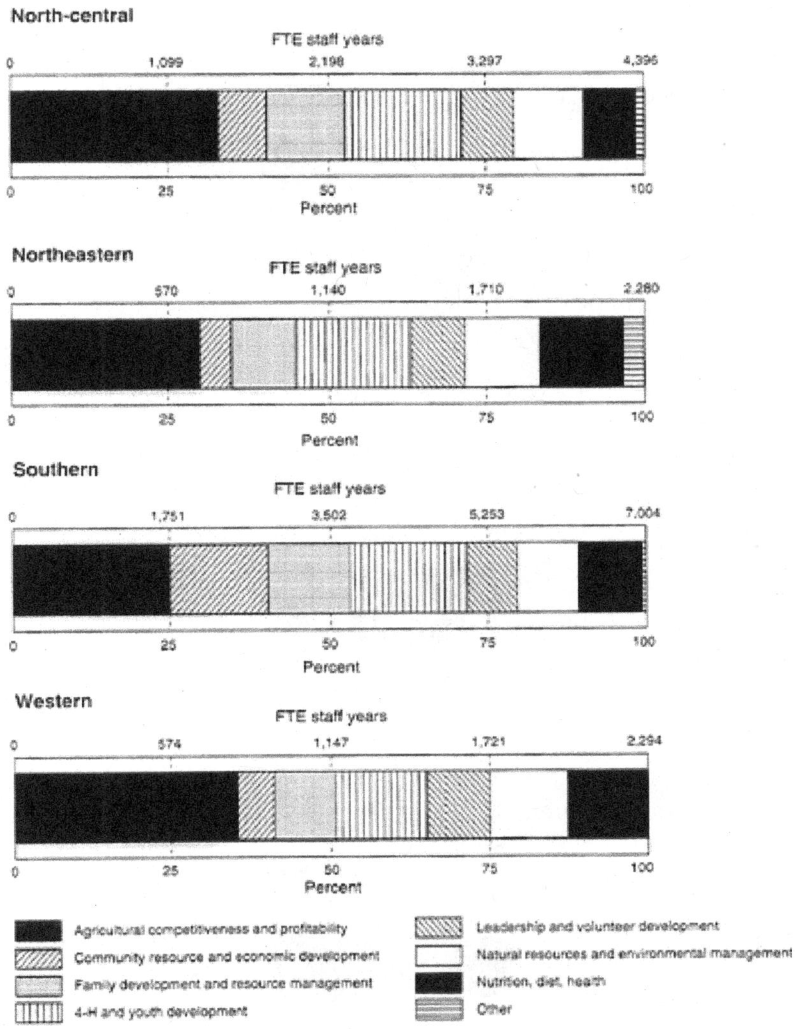


**FIGURE 5-2**

*During 1992 extension staff divided their time (full-time equivalent [FTE] staff years) among base programs involving farm service, community development, natural resource management, and consumer education programs.*

- Extension specialists are located at every land grant college of agriculture, and extension agents operate in almost every county in the nation. Figure 5-2 shows the distribution of each region's extension staff among Cooperative Extension's seven base program areas. Extension staff are located mostly in the south (Texas has by far the largest program) and in the north-central region of the country (there staff are more evenly distributed across states). About 30 percent of extension staff are located in the west and northeast in approximately equal numbers (see Appendix Table 3 for state-by-state extension staff allocations). The regional distribution of extension staff roughly mirrors the regional distribution of the nation's rural population (Figure 5-3).

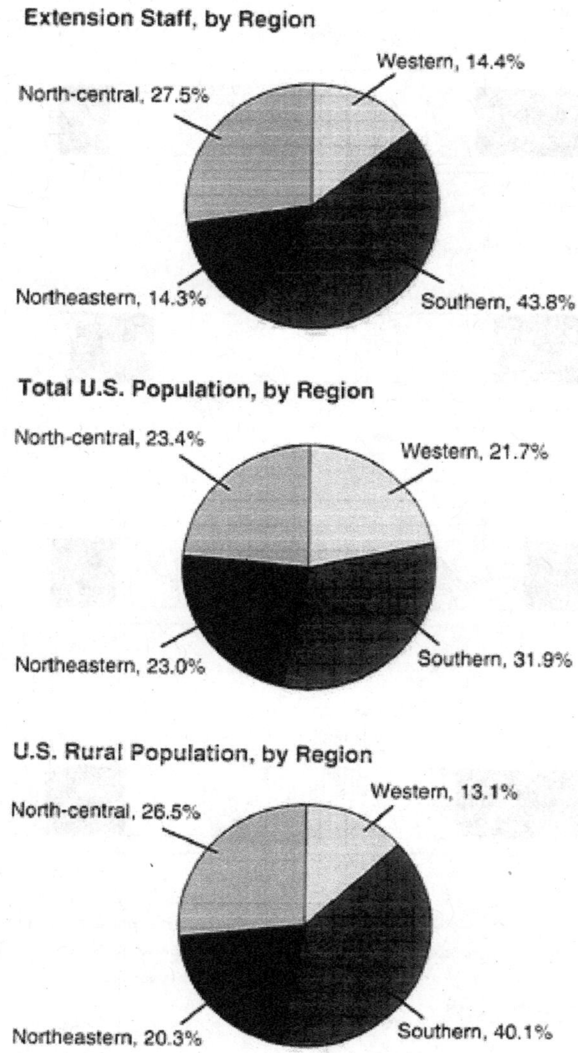


**FIGURE 5-3**
*In 1992 the largest number of extension staff was in the southern region. The geographic allocation of extension staff relates closely to the distribution of the nation's rural population.*

- Agricultural competitiveness and profitability were the goals of the largest base program in 1992, accounting for about one-third of all extension staff years. However, the efforts of 45 percent of extension staff were targeted toward the four related goals of community development, family development, youth development, and leadership development—programs that may be applicable in both rural and urban areas (Figure 5-4).

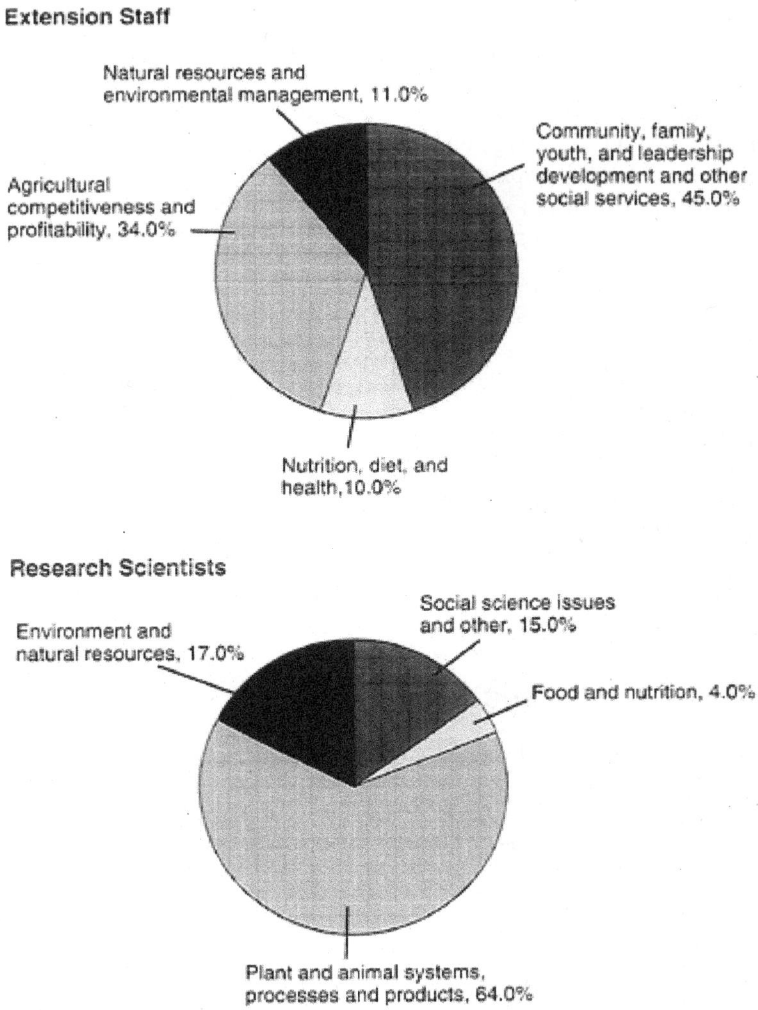


**FIGURE 5-4**
*Charts show national allocation, by program area, of extension staff and SAES research scientists for 1992.*

- The averages cited above belie some pronounced differences across states. For example, California, with a highly developed commercial agriculture, devotes more than one-half of its extension staff years to "agricultural competitiveness and profitability"; while West Virginia, with a larger low-income population, allocates two-thirds of its extension staff to programs aimed at the development of communities, families, youth, and leadership (Appendix Table 3).

- Extension staff—both specialists at the colleges and agents at the county level—often draw on information generated by the research of SAES scientists. The allocation of extension staff and research scientists among program areas is, however, quite different. Of the research scientists' time, 64 percent was allocated in 1992 to research that could directly benefit farm productivity (such as research on plant and animal systems) and the sales of farm products (such as research on "processing for value added"). Only 13 percent was directly targeted toward social science issues (which would include rural, community, and leadership development) (Figure 5-4).
- Similarly, 10 percent of extension staff was assigned to nutrition, diet and health programs, in contrast to the 4 percent of research scientists that reported research in these areas (Figure 5-4).

## ISSUES FOR DISCUSSION

- Does the research-extension continuum, for which the land grant system is so well known, still function well? Is extension responding to a different set of national, state, and local needs than is college-based agricultural research?
- Would an expansion of nutrition and social science research provide a sounder base for extension activities in nutrition education and community and rural development?
- What is the role of independent private crop consultants and agricultural input firms vis-à-vis public extension services?
- What indicators might be developed to measure the benefits of public investments in extension programs?

## SUGGESTED READINGS

U.S. Department of Agriculture, Extension Service. *Cooperative Extension Roles and Relationships for a New Era.* Washington, D.C.: U.S. Department of Agriculture, 1990.

Feller, Irwin, et al. *Agricultural Technology Delivery System*, 5 vols. University Park: Pennsylvania State University, 1984.

Natural Resources and Environment Division, Economic Research Service, U.S. Department of Agriculture. *AREI Updates: Crop Consultants*, No. 3. Washington, D.C.: U.S. Department of Agriculture, 1995.

Rasmussen, Wayne D. *Taking the University to the People: Seventy-Five Years of Cooperative Extension.* Ames:Iowa State University Press, 1989.

# 6

# THE SHIFTING BASE OF FINANCIAL SUPPORT FOR LAND GRANT COLLEGE RESEARCH AND EXTENSION

*This chapter revisits the college research and extension programs from a different perspective—their changing base of support from federal, state, local government, and private entities. It discusses the role of non-USDA federal agencies in supporting agricultural research; and the changing mechanisms for funding, including traditional formula-based funds, competitive grants, and grants "earmarked" by Congress, are described and compared.*

- Land grant colleges of agriculture are public institutions supported by the revenues generated by U.S. federal, state, and local governments. The public has, therefore, a stake in the accomplishments and services of the system. Public input into research, education, and extension direction and priorities can occur in a number of ways, but its impact may be limited or require significant time to result in redirection. Over time, changes in the respective roles of public and private entities that provide college funding, as well as changes in the mechanisms used for channeling public funds to the system, occur. These changes have an effect on the colleges' activities and priorities and the public's opportunities to influence them. Changes in the distribution of funding across institutions in the system—or differences in funding mechanisms across institutions—can affect linkages and cooperation within the system and each college's role in the integrated whole.
- The U.S. public has a $3 billion stake in the combined research and extension activities of the land grant colleges of agriculture and their forestry and veterinary medicine counterparts. In 1992 state budgets funded slightly less than one-half of all combined research and extension expenditures at these institutions. In the same year, federal funding equaled approximately one-third of research expenditures and between one-quarter and one-third of cooperative extension costs (Table 6-1).
- Private funds are, however, of increasing importance to the colleges' financial status and the shaping of their research programs. Over the last 20 years private funds for research at SAESs, which include grants from industry and nonprofit organizations and revenue generated by commercial sales of products (such as college-owned livestock and livestock products) and licenses, grew faster than either federal or state support. These nonpublic sources now fund 19 percent of research expenditures by all SAESs and colleges reporting to CRIS (Table 6-1). Grants from industry, often funds generated by the "check-off" programs of commodity groups (such as those for beef, pork, soybeans, and wheat), make up about 40 percent of these nonpublic funds.

The Structure of federal support for research at colleges of agriculture is shifting away from fixed formulas toward competitive grands based on scientific merit and special grands earmarked by Congress.

- Meanwhile, support for cooperative extension activities has shifted since 1972 from the federal to the state and local governments (Table 6-1). In general, in the last 20 years there has been stronger federal support for the colleges' research programs than for their extension activities. However, lack of complete data on private sector activities in research and extension make it difficult to say with certainty which activity has become less public and more private.
- Federal dollars for research conducted at land grant colleges of agriculture flow to the colleges through four funding mechanisms:

  1. formula-based grants administered by USDA,
  2. special grants earmarked by Congress for specific institutions and administered by USDA,
  3. competitive grants awarded and administered by USDA, and
  4. other research grants (or cooperative agreements) awarded by other federal agencies (including some USDA agencies not responsible for administering the grants in the first three categories).

TABLE 6-1 Sources of Support for Research and Extension Activities at the 1862 and 1890 Institutions and Related Colleges and Schools of Forestry and Veterinary Medicine, 1972–1992

| | Research[a] | | | | Extension[b] | | | |
|---|---|---|---|---|---|---|---|---|
| Year | Federal | State | Private | Total | Federal | State | Local | Total |
| *Millions of Dollars* | | | | | | | | |
| 1972 | 118 | 205 | 51 | 374 | 149 | 136 | 70 | 354 |
| 1977 | 201 | 341 | 94 | 636 | 198 | 220 | 105 | 524 |
| 1982 | 355 | 544 | 169 | 1,069 | 302 | 368 | 182 | 852 |
| 1987 | 415 | 778 | 253 | 1,447 | 319 | 500 | 229 | 1,048 |
| 1992 | 631 | 981 | 380 | 1,992 | 401 | 652 | 333 | 1,389 |
| *Average Annual Growth (percent)* | | | | | | | | |
| 1972–1977 | 14 | 13 | 17 | 14 | 7 | 12 | 10 | 9 |
| 1977–1982 | 15 | 12 | 16 | 14 | 11 | 13 | 15 | 13 |
| 1982–1987 | 3 | 9 | 10 | 7 | 1 | 7 | 5 | 5 |
| 1987–1992 | 10 | 5 | 10 | 8 | 5 | 6 | 9 | 6 |
| 1972–1992 | 21 | 19 | 32 | 22 | 8 | 19 | 19 | 15 |

NOTE: Private funds for research include grants from industry and nonprofit organizations and from the sale of products and licenses.
[a] Research funds are expenditures reported in CRIS.
[b] Extension funds are budget appropriations reported by the Extension Service.
SOURCE: Data are from USDA Current Research Information System (CRIS) and USDA Extension Service (ERS).

TABLE 6-2 Sources of Federal Funds (thousands of dollars) to the 1862 and 1890 Institutions and Related Colleges and Schools of Forestry and Veterinary Medicine, 1972–1992

| | Funding Mechanism | | | | |
|---|---|---|---|---|---|
| Fiscal Year | Formula Funds | Special Research Grants | Competitive Research Grants | Other Federal Funds | Total Federal Funds |
| *1862 State Agricultural Experiment Stations* | | | | | |
| 1992 | 177,459 (33) | 76,742 (14) | 55,745 (10) | 226,037 (42) | 535,983 |
| 1987 | 153,727 (44) | 27,813 (8) | 22,751 (6) | 147,925 (42) | 352,216 |
| 1982 | 147,775 (49) | 20,726 (7) | 10,452 (3) | 123,352 (41) | 302,305 |
| 1977 | 100,223 (55) | 8,439 (5) | NA | 70,793 (39) | 180,656 |
| 1972 | 67,502 (62) | 3,617 (3) | NA | 36,861 (34) | 108,033 |
| *1890 Colleges and Universities* | | | | | |
| 1992 | 25,823 (90) | 2,643 (9) | 111 (<1) | 260 (1) | 28,837 |
| 1987 | 20,460 (99) | 90 (<1) | NA | 220 (<1) | 20,770 |
| 1982 | 19,254 (100) | 0 (0) | NA | 18 (<1) | 19,272 |
| 1977 | NA | 13,130 (99) | NA | 153 (1) | 13,283 |
| 1972 | NA | 8,883 (100) | NA | 0 (0) | 8,883 |
| *Forestry Schools* | | | | | |
| 1992 | 4,624 (21) | 1,357 (6) | 2,179 (10) | 13,750 (63) | 21,910 |
| 1987 | 2,699 (26) | 193 (2) | 931 (9) | 6,535 (63) | 10,358 |
| 1982 | 2,472 (24) | 80 (1) | NA | 7,726 (75) | 10,278 |
| 1977 | 2,033 (29) | 309 (4) | NA | 4,674 (67) | 7,016 |
| 1972 | 900 (47) | 0 (0) | NA | 1,024 (53) | 1,923 |
| *Schools of Veterinary Medicine* | | | | | |
| 1992 | 1,493 (3) | 187 (<1) | 2,829 (6) | 39,586 (90) | 44,095 |
| 1987 | 1,405 (4) | 1,131 (4) | 310 (1) | 28,638 (91) | 31,485 |
| 1982 | 1,214 (4) | 1,047 (4) | NA | 26,911 (92) | 29,172 |
| 1977 | NA | NA | NA | NA | NA |
| 1972 | NA | NA | NA | NA | NA |
| *All Above Institutions* | | | | | |
| 1992 | 209,400 (33) | 80,929 (13) | 60,863 (10) | 279,634 (44) | 630,825 |
| 1987 | 178,291 (43) | 29,227 (7) | 23,992 (6) | 183,318 (44) | 414,829 |
| 1982 | 170,715 (47) | 21,853 (6) | 10,452 (3) | 158,007 (44) | 361,028 |
| 1977 | 102,256 (51) | 21,879 (11) | NA | 75,619 (38) | 200,955 |
| 1972 | 68,402 (58) | 12,500 (11) | NA | 37,884 (32) | 118,839 |

NOTE: Formula funds are administered by CSRS based on funding legislation sponsored by Hatch, McIntire-Stennis, and Evans-Allen and on animal health and disease programs. Other federal funds are contributed by non-CSRS federal agencies. Figures are expenditures of funds reported by the institutions themselves. Number in parentheses is percent of total federal funds. NA, data not available because either program is not active or institution type does not report.

SOURCE: Data are from USDA Current Research Information System (CRIS).

- Between 1935 (when the Bankhead-Jones Act was passed) and the late 1980s, formula funding, as established by the act, comprised the largest category of federal monies for SAES research (Table 6-2). Hatch funds are funneled only to SAESs and their allocation among these institutions is inflexible with respect to the focus, scientific review, or outcome of the station's research (see box copy, p. 78). System-wide priorities for these (and other agricultural research) funds are laid out by ESCOP, but in reality each SAES has wide latitude in deciding how to allocate and use formula funds.

## FORMULA FUNDING MECHANISMS: PAYMENTS TO AGRICULTURAL EXPERIMENT STATIONS UNDER THE HATCH ACT

Funds received as a result of the Hatch Act (first enacted in 1887) are allocated for research to promote sound and prosperous agriculture and rural life to the state agricultural experiment stations (SAESs) of the 50 states, the District of Columbia, Puerto Rico, Guam, the Virgin Islands, Micronesia, American Samoa, and the Northern Marianas Islands. The Amended Hatch Act (1955) provides that the distribution of federal payments to states for FY 1955 shall become a fixed base and that any sums appropriated in excess of the 1955 amount shall be distributed in the following manner:

- 20 percent allotted equally to each state;
- not less than 52 percent allotted to the states as follows:

- one-half in an amount proportionate to each state's share of the total U.S. rural population, and
- one-half in an amount proportionate to each state's share of the total U.S. farm population,

- not more than 25 percent shall be allotted to the states for cooperative regional research in which two or more SAESs are cooperating to solve problems that concern the agriculture of more than one state; and
- 3 percent shall be available to the Secretary of Agriculture for the administration of the act.

The Hatch Act also provides that any amount in excess of $90,000 available to any state, exclusive of the regional research fund, shall be matched by the state out of its own funds available for research and for the establishment and maintenance of facilities necessary for the performance of such research. In the case of Guam, the Virgin Islands, Micronesia, American Samoa, and the Northern Marianas Islands, agencies are required by law to waive any requirement for local matching funds for federal formula funds less than $200,000.

Three percent of funds appropriated under the Hatch Act is set aside for federal administration, which includes disbursement of funds and a continuous review and evaluation of the research programs of the state agricultural experiment stations supported wholly or in part by Hatch Act funds. USDA's Cooperative State Research Service (now merged with the Extension Service to form the Cooperative State Research, Education, and Extension Service or CREES) encourages and assists in establishing research linkages and partnerships within and between the states and actively participates in the planning and coordination of research programs between the states and USDA at the regional and national levels.

SOURCE: National Research Council. 1989. Investing in Research: A Proposal to Strengthen the Agricultural, Food, and Environmental System. Washington, D.C.: National Academy Press.

- Three other formula-based grant programs are aimed at forestry research (McIntire-Stennis funds), animal-health research, and the programs of the 1890 colleges (Evans-Allen funds). McIntire Stennis funds flow to both forestry schools and SAESs; animal-health formula funds go to both veterinary medicine colleges and SAESs.
- Between 1987 and 1992 the amount "other federal funds" surpassed the amount of formula (or Hatch) funds. The system as a whole (and in particular the SAESs at the 1862s) has reduced its reliance on the traditional formula funds and has diversified its funding portfolio by participating in the grants programs (typically competitive grants programs) of other federal agencies. However, researchers at the 1890s schools are still overwhelmingly dependent on USDA-administered formula funds.
- USDA-administered competitive grants, though still small currently, also increased in importance—about sixfold between 1982 and 1992. The National Research Initiative Competitive Grants Program (NRI)—the main competitive grants program administered by USDA—is accessible to scientists at all public and private universities. In fact, about 27 percent of NRI grant applications have not come from land grant universities (National Research Council, 1994). Thus while land grant colleges of agriculture are broadening their forms of support, they must also compete more actively with non-land grant schools for research funds.
- Access to senators and congressmen, and the influence of those politicians on appropriations for agricultural research, has been of growing importance to land grant colleges of agriculture. Much more significant today than they were 20 years ago are the special research funds earmarked by Congress (see Table 6-2). They now account for a larger percentage of total SAES research expenditures than do USDA's competitive grants (Figure 6-1).

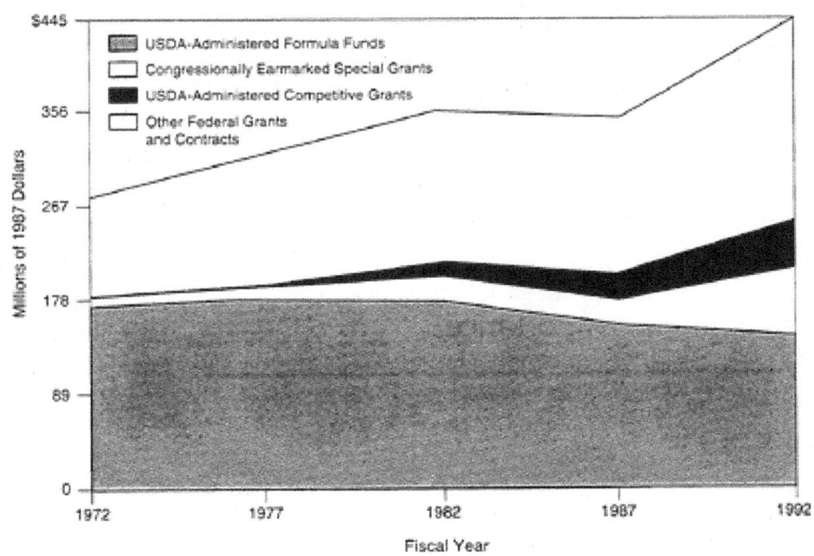


**FIGURE 6-1**

*Over time, USDA-administered formula funds have decreased in importance in relation to other sources of research expenditures at 1862 state agricultural experiment stations.*

Excluding USDA, the National Institute of Health, the Agency for International Development, and the National Science Foundation are the largest federal supporters of agricultural research.

- Table 6-3 ranks 1862 colleges of agriculture by receipt of special research grants earmarked for them by Congress. At some colleges research funding appears to benefit from the college's political access and influence. In 1992 special research grants accounted for 25 percent of SAES research funding at the University of Vermont, 13 percent at the University of Hawaii, 11 percent at Michigan and Iowa State universities, and 10 percent at Mississippi State, North Dakota, and New Mexico.
- Federal agencies other than USDA that provide the most support for agricultural research include the nation's major science funding agencies—National Institutes of Health (NIH) and the National Science Foundation (NSF)—and the U.S. Agency for International Development (AID) (Table 6-4). AID funds are directed to international agricultural research, particularly toward collaborative research support programs involving AID, U.S. universities, and host developing country institutions (CRSPs), which depend heavily on AID for their support.
- In recent years, two other federal agencies, the U.S. Department of Health and Human Services and the U.S. Department of Energy (DOE), have gained prominence in agricultural research funding. SAES research expenditures based on grants from these two agencies approximately doubled between 1987 and 1992 (Table 6-4). Also, DOE collaborates with USDA and NSF in supporting grants for plant biology research (National Science Foundation, 1993).

State and federal funds are still the financial mainstay of the land grant college of agriculture system, although private and local partnerships are increasing in importance.

TABLE 6-3 Special Research Grants Allocated to 1862 Institutions, Ranked by Amount Received, 1992

| Institution | Amount (thousands) | Percent of Experiment Station Research Expenditures |
|---|---|---|
| Michigan State U. | $6,573 | 11 |
| Iowa State U. | 6,485 | 11 |
| U. of California | 3,905 | 3 |
| Mississippi State U. | 3,543 | 10 |
| U. of Hawaii | 3,061 | 13 |
| U. of Nebraska | 2,914 | 6 |
| Purdue U. | 2,887 | 6 |
| U. of Arkansas | 2,789 | 9 |
| Oregon State U. | 2,774 | 7 |
| Louisiana State U. | 2,587 | 7 |
| Washington State U. | 2,527 | 7 |
| North Dakota State U. | 2,511 | 10 |
| Colorado State U. | 2,197 | 7 |
| Pennsylvania State U. | 2,140 | 6 |
| U. of Florida | 2,036 | 2 |
| Cornell U. | 1,877 | 3 |
| Texas A&M U. | 1,835 | 2 |
| U. of Vermont | 1,814 | 25 |
| U. of Georgia | 1,768 | 4 |
| U. of Missouri | 1,699 | 5 |
| U. of Minnesota | 1,530 | 3 |
| Ohio State U. | 1,411 | 4 |
| Kansas State U. | 1,383 | 3 |
| Rutgers—The State U., Cook College | 1,324 | 5 |
| New Mexico State U. | 1,243 | 10 |
| U. of Idaho | 1,000 | 5 |
| Oklahoma State U. | 862 | 3 |
| Clemson U. | 837 | 3 |
| Auburn U. | 799 | 2 |
| U. of Wisconsin | 770 | 1 |
| U. of Illinois | 733 | 2 |
| North Carolina State U. | 708 | 1 |
| U. of Maine | 673 | 5 |
| Geneva AES | 659 | 6 |
| U. of Kentucky | 597 | 2 |
| U. of Maryland | 568 | 3 |
| U. of Massachusetts | 485 | 4 |
| Montana State U. | 468 | 3 |
| U. of Connecticut | 378 | 6 |
| U. of Arizona | 378 | 1 |
| Virginia Polytechnic Institute and State U. | 358 | 1 |
| U. of Puerto Rico | 332 | 3 |
| U. of Guam | 279 | 10 |
| U. of the Virgin Islands | 214 | 15 |
| U. of Rhode Island | 166 | 5 |
| Utah State U. | 138 | 1 |
| U. of Tennessee | 138 | 1 |
| New Haven AES | 114 | 2 |
| U. of New Hampshire | 101 | 2 |
| South Dakota State U. | 82 | 1 |
| West Virginia U. | 45 | 1 |
| U. of Wyoming | 29 | <1 |
| U. of Alaska | 5 | <1 |
| U. of Delaware | 5 | <1 |
| U. of Nevada | 5 | <1 |
| Total | 76,742 | 5 |

SOURCE: Data are from USDA Current Research Information System (CRIS).

TABLE 6-4 Sources of Other Federal Funds (nominal dollars in thousands) Received by 1862 SAESs, 1972–1992

| | Funding Agency | | | | | | | | | | | |
|---|---|---|---|---|---|---|---|---|---|---|---|---|
| Year | USDA-CGCA | NIH | NSF | AID | HHS | PHS | DOD | DOE | NASA | TVA | Other | Total |
| 1992 | 53,849 | 34,166 | 24,601 | 27,771 | 13,148 | 10,152 | 3,847 | 9,190 | 4,303 | 1,242 | 43,767 | 226,037 |
| 1987 | 33,018 | 18,251 | 18,996 | 21,587 | 6,616 | 15,183 | 3,906 | 4,850 | 2,483 | 692 | 22,343 | 147,925 |
| 1982 | 30,998 | 10,529 | 15,205 | 14,141 | 7,467 | 8,602 | 1,831 | 4,244 | 2,230 | 443 | 27,662 | 123,352 |
| 1977 | 11,739 | 10,439 | 10,559 | 7,620 | 3,022 | 6,742 | 1,085 | 2,103 | 1,613 | 136 | 15,736 | 70,793 |
| 1972 | 6,850 | 5,801 | 4,502 | 2,712 | 1,233 | 7,420 | 1,029 | 785 | 902 | 95 | 5,531 | 36,861 |

Abbreviations: USDA-CGCA, USDA Contracts, Grants, and Cooperative Agreements; NIH, National Institutes of Health; NSF, National Science Foundation; AID, U.S. Agency for International Development; HHS, U.S. Department of Health and Human Services; PHS, U.S. Public Health Service; DOD, U.S. Department of Defense; DOE, U.S. Department of Energy; NASA, National Aeronautics and Space Administration; TVA, Tennessee Valley Authority.
SOURCE: Data are from USDA Current Research Information System (CRIS).

## ISSUES FOR DISCUSSION

- Over the decades, state support relative to federal support of the land grant system has increased. More recently, the role of private sources of funds, such as industry grants, has also increased in relative importance. What is the continued role for federal funding of the activities of colleges of agriculture? Are there issues and problems of national concern and scope to which the states may not direct adequate resources?
- How much and what types of agricultural research would the private sector conduct in collaboration with colleges of agriculture? Will the growing use of private funds for college research help or hinder research on long-term projects of broad national interest?
- The role of formula grants is decreasing, particularly at the large research universities, while competitive grants and congressionally earmarked grants are increasing. Should competitive grants compose a larger share of agricultural research funding, and what would be the implications for the distribution and use of funds in the system? If legislators were proposing a formula-based funding mechanism today, against the context of today's state economies, how might it differ from the one proposed many decades ago?
- Some colleges of agriculture received substantial portions of their federal funds from non-USDA federal agencies. What should be the role of these other agencies, in relation to USDA's, in funding and influencing agricultural research priorities at colleges of agriculture and other institutions?

## SUGGESTED READINGS

National Research Council. *Investing in Research: A Proposal to Strengthen the Agricultural, Food, and Environmental System*. Washington, D.C.: National Academy Press, 1989.

National Research Council. *Investing in the National Research Initiative: An Update of the Competitive Grants Program in the U.S. Department of Agriculture*. Washington, D.C.: National Academy Press, 1994.

National Research Council. *Toward Sustainability: A Plan for Collaborative Research on Agriculture and Natural Resource Management*. Washington, D.C.: National Academy Press, 1991.

Ruttan, Vernon W., and Carl E. Pray, eds. *Policy for Agricultural Research* . Westview, CT: Westview Press, 1987.

U.S. Department of Agriculture, Cooperative State Research Service. *Dynamics of the Research Investment: Issues and Trends in the Agricultural Research System*. Washington, D.C.: U.S. Department of Agriculture, 1993.

Von Braun, Joachim, Raymond Hopkins, Detlev Puetz, and Rajul Pandya-Lorch. *Aid to Agriculture: Reversing the Decline*. Washington, D.C.: International Food Policy Research Institute, 1993.

# 7

# PROFILES OF THE LAND GRANT COLLEGES OF AGRICULTURE: Comparisons of Structure, Focus, and Funding

*This chapter takes a more disaggregated view of the college of agriculture research system. Its purpose is to allow the reader to see the land grant agricultural research system as a network of individual institutions with important similarities and differences. The chapter compares college names, administrative and organizational structures, emphases of research programs, and approaches to funding research. The system-wide organization of livestock and crops research is also explored; one reason is to determine whether institutions in close geographical proximity share research interests that relate to their region's farm economy.*

## PROFILES-PART A THE COLLEGES' CHANGING NAMES

- Thirty years ago all agriculture colleges at 1862 institutions were either "colleges of agriculture" or "colleges of agriculture and home economics" (Table 7-1). Today, fewer than one-half have retained these names. At a some institutions, home economics departments are now separate colleges with names such as "College of Applied Human Sciences" at Colorado State U., "College of Human Resources" at the University of Delaware, and "College of Human Ecology" at Kansas State U.
- After "college of agriculture" or "college of agricultural sciences," the most common name is "college of agriculture and life sciences," reflecting an increased orientation toward the basic sciences. Since 1988, the most popular new college name, as resource and environmental issues have gained prominence, has been "college of agriculture and natural resources." Another fast-growing category is names without "agriculture" in the title at all. This increasing diversity of names of colleges of agriculture is one sign of change in the land grant college system.

Fewer than one-half of the 1862 colleges retain the name "College of Agriculture" or "College of Agricultural Sciences."

TABLE 7-1 Changing Names of 1862 Colleges (percent)

| | Year | | | |
|---|---|---|---|---|
| Name | 1962 | 1974 | 1988 | 1993 |
| Agriculture[a] | 86 | 64 | 58 | 45 |
| and Home Economics | 14 | 8 | 8 | 7 |
| and Natural Resources | 0 | 6 | 8 | 13 |
| and Life Sciences | 0 | 14 | 14 | 15 |
| and Environment | 0 | 4 | 2 | 4 |
| "Agriculture" not part of title | 0 | 2 | 6 | 9 |
| Other | 0 | 2 | 4 | 7 |

[a] Of 25 colleges called either "College of Agriculture" or "College of Agricultural Sciences," 7 (28 percent) are called the latter.
SOURCES: Data for 1962, 1974, and 1988 are from Myers, J. H. 1991. Rethinking the Outlook of Colleges Whose Roots Have Been in Agriculture. Davis: University of California. Data for 1993 are from USDA Food and Agricultural Education Information System (FAEIS).

Despite differences in college names, administration, and organization, the majority of faculty and staff at colleges of agriculture work in academic departments with a production-agriculture focus.

- Veterinary medicine, forestry, and home economics are sometimes programs within the college of agriculture and sometimes separate administrative units. Some colleges have interdisciplinary centers and others are moving toward multidisciplinary clustering of departments. Cooperative extension functions are sometimes administered by the dean of the college of agriculture (who typically administers the experiment station) and sometimes by another university administrator. (For example, at the University of Wisconsin, extension is a separate "campus" with university-wide functions.) For some colleges of agriculture, the office of academic affairs is under the administration of a university-wide official (see box copy, p. 85).
- Despite differences in college organization and administration, at most 1862 colleges a large number of faculty and staff work in academic areas that, by department name, imply a production-agriculture focus —for example, agricultural engineering, agronomy and soil science, animal sciences, entomology, plant pathology, and other specific plant sciences. In a 1993 survey of resident-instruction faculty in land grant colleges of agriculture, natural resources, and forestry, it was estimated that

- about 47 percent of faculty taught general agriculture, animal sciences, plant science, soil sciences, or agricultural engineering/mechanization;
- less than 19 percent taught agricultural business and management (including agricultural economics) and education, communication, and social sciences;
- 17 percent taught natural resources and forest sciences;
- 11 percent taught related biological or physical sciences; and
- slightly more than 5 percent taught food science and human nutrition (Table 7-2).

Including faculty and staff without resident-instruction appointments would put an even higher percentage in production-agriculture fields.

## FIVE PROFILES

Colleges administer somewhat differently the three functions of teaching, research, and extension, as differently as they configure their academic programs and departments. Five profiles are given here—four colleges and one university agriculture system—to illustrate the variety.

**ADMINISTRATIVE STRUCTURES AND DEPARTMENTAL CONFIGURATION AT FIVE SELECTED 1862 "COLLEGES OF AGRICULTURE"**

**University of California**

The university's Division of Agriculture and Natural Resources is the university system's administrative umbrella for the agricultural experiment station, cooperative extension, and the College of Natural Resources at Berkeley; the College of Agriculture and Environmental Sciences and the School of Veterinary Medicine at Davis; and the College of Natural and Agricultural Sciences at Riverside.

The division is directed by a university vice president who also directs the experiment station and cooperative extension. Each of the three campus-based colleges has a dean who also serves as an associate director of the experiment station. There are also four regional program directors who oversee regional research and extension programs of the state-wide field offices.

***Berkeley***

Berkeley's College of Natural Resources has melded four of its seven departments—plant pathology, soil science, entomology, and forestry—into one department of environmental science, policy, and management. Other departments include agricultural and resource economics, nutritional sciences, and plant biology.

***Davis***

Davis' College of Agriculture and Environmental Sciences is the largest of the three campus-based colleges and is self-contained, encompassing traditional production agriculture departments in plant sciences and animal biology, human health and development departments, and environmental and natural resource science and policy departments. Faculty of the Division of Biological Sciences also have experiment station appointments, making for strong links to basic science.

***Riverside***

Riverside's College of Natural and Agricultural Resources includes biochemistry, biology, botany, earth sciences, entomology, nematology, plant pathology, soil and environmental sciences, and statistics.

**University of Connecticut**

The University of Connecticut's College of Agriculture and Natural Resources is headed by a dean who also directs the college's experiment station and cooperative extension. There are three associate deans, one each administering the experiment station, cooperative extension, and the college. The college has departments of agricultural and resource economics, animal science, natural resources management and engineering, nutritional sciences, pathobiology, and plant science. It also has a school of family studies; interdisciplinary centers for environmental health, food marketing policy, and wildlife disease; and an institute of water resources (staffed by faculty and staff of disciplinary departments).

The state also has a second experiment station, located in New Haven, not affiliated with a university and with a separate administration. Scientists have station research appointments only—in chemistry, biochemistry and genetics, entomology, forestry and horticulture, plant pathology and ecology, and soil and water.

**University of Missouri**

The University of Missouri's College of Agriculture, Food, and Natural Resources is headed by a dean who also directs the experiment station. There are also colleges of human environmental sciences and veterinary medicine. Assistant directors of the experiment station head these latter units. A university vice provost oversees extension, and an associate dean of the college is extension's associate director. The college has an associate dean as the administrative head of academic programs.

Separate departments specific to each discipline have been largely eliminated. Experiment station researchers and extension specialists are clustered in several large units: agricultural information, animal science, biochemistry, food science and engineering, the school of natural resources, plant sciences, and social sciences.

**New Mexico State University**

New Mexico State U.'s College of Agriculture and Home Economics has a dean who also serves as the chief administrative officer. There are three associate deans who serve as, respectively, the director of the experiment station, the director of cooperative extension, and the director of academic programs.

Departments include agricultural economics and business; agronomy and horticulture; animal and range sciences; entomology, plant pathology, and weed science; experimental statistics; fishery and wildlife sciences; and home economics. There is also a school for hospitality and tourism services, a plant genetic engineering laboratory, and a number of off-campus agricultural science centers.

**North Carolina State University**

North Carolina State University's (NSCU's) College of Agriculture and Life Sciences (CALS) has three divisions: academic programs, research, and extension. CALS's North Carolina Agricultural Research Service (NCARS) is the state's primary agency for research in agriculture, life sciences, forestry, and home economics; it is also North Carolina's agricultural experiment station. NCARS research is conducted in CALS and in the colleges of Forest Resources and

Veterinary Medicine at NCSU and in the School of Human Environmental Sciences at U. North Carolina at Greensboro.

Within CALS, NCARS coordinates research in 19 academic departments, working in partnership with extension and teaching. Departments include agricultural communications, animal science, biochemistry, botany, crop science, economics and business, engineering, entomology, food science, horticultural science, human environmental sciences, microbiology, plant pathology, poultry science, sociology and anthropology, soil science, statistics, toxicology, and zoology.

**STAFF PROFILE**

The table below shows the number of staff with doctorate degrees, by appointment type, at the five colleges. At all five, the most common appointment type combines research and teaching. Among these five colleges, research-only appointments are common only at the U. of Connecticut, largely because of the presence of Connecticut's second state experiment station located away from the university; but the U. of Connecticut also has the highest portion of three-way appointments—extension, teaching, and research. Extension-only appointments are more common at New Mexico State U. than in the other four states.

**Number of College and Professional Staff Holding Doctorate Degrees, by Appointment Type, 1993-1994**

| University | Research Only | Research/ Teaching | Research/ Extension | Extension Only | Research/ Teaching/ Extension |
|---|---|---|---|---|---|
| UC Davis* | 13 | 458 | 31 | 33 | 2 |
| UCONN† | 38 | 42 | 3 | 4 | 34 |
| UMO | 4 | 192 | 18 | 16 | 39 |
| NMSU | 10 | 79 | 5 | 14 | 8 |
| NCSU‡ | 90 | 345 | 57 | 21 | 77 |

NOTE: Drawn from 1993–94 Directory of Professional Workers in State Agricultural Experiment Stations and Other Cooperating State Institutions, these numbers should be taken only as very rough estimates of number and distribution of Ph.D. and D.V.M. staff. They are not converted to full-time equivalents. There may also be double counting because some faculty have appointments in more than one department or unit. Faculty on leave and emeritus faculty are included. The assumption is that Ph.D. faculty members who have experiment station appointments have a research function; if a college appointment, then teaching functions; if an extension appointment, then an extension function.

* Includes faculty of Division of Biological Sciences and School of Veterinary Medicine listed in the Directory. However, many of the veterinary school faculty do not have experiment station appointments and are thus not counted.

† Includes staff of New Haven Agricultural Experiment Station, which is administered separately from the experiment station at UCONN. Most staff with research only appointments are associated with the New Haven Station.

‡ Includes Ph.D. staff of College of Forest Resources, the majority of whom have experiment station appointments.

TABLE 7-2 Number of Resident Instruction Faculty in Land Grant Colleges of Agriculture, Natural Resources, and Forestry by Academic Rank, 1993

| Academic Rank | General Agriculture | Animal Science | Plant Science | Soil Science | Agricultural Business and Management | Education, Communication, Soil Sciences |
|---|---|---|---|---|---|---|
| Professor | 31 | 446 | 602 | 148 | 568 | 105 |
| Associate professor | 11 | 246 | 374 | 66 | 175 | 77 |
| Assistant professor | 5 | 149 | 232 | 34 | 88 | 42 |
| Instructor | 9 | 22 | 21 | 0 | 13 | 5 |
| Totals | 56 | 863 | 1,229 | 248 | 844 | 229 |

SOURCE: Data are from USDA Food and Agricultural Education Information System (FAEIS).

- Schools such as the University of Kentucky, which is located in a state where 6 percent of jobs are in farm production and 21 percent are in farm-related industries, are traditionally production oriented. At Kentucky's college of agriculture

- 200-plus nonadministrative professional staff (with teaching, research, or extension appointments) are in departments of agronomy, animal science, agricultural engineering, entomology, horticulture, and plant pathology;
- 17 are in forestry;
- 30 are in the college of human environmental sciences;
- 49 are in agricultural economics and sociology; and
- 42 in veterinary sciences (U.S. Department of Agriculture, 1993b).

- At the University of Connecticut, located in a state where less than 0.5 percent of jobs are on farms and only 11 percent are in farm-related business, there is less emphasis on production agriculture and staff are far fewer and configured differently:

- 53 nonadministrative staff are in animal sciences, plant sciences, and pathobiology;
- 29 are in the school of family studies and the department of nutritional sciences;
- 11 are in natural resources management and engineering; and
- 11 are in agricultural and resource economics (U.S. Department of Agriculture, 1993b).

The structure of staff responsibility for teaching, research, and extension also differs across the system, but the most common appointment type combines research and teaching.

| Academic Rank | Natural Resources | Forest Sciences | Food Science/ Human Nutrition | Agricultural Engineering/ Mechanization | Related Biological/ Physical Sciences | Total Across Area |
|---|---|---|---|---|---|---|
| Professor | 254 | 225 | 143 | 177 | 373 | 3,072 |
| Associate Professor | 143 | 155 | 91 | 81 | 155 | 1,574 |
| Assistant Professor | 94 | 83 | 60 | 70 | 91 | 948 |
| Instructor | 13 | 17 | 8 | 11 | 11 | 130 |
| Totals | 504 | 480 | 302 | 339 | 630 | 5,724 |

TABLE 7-3 Agriculture and Natural Resources and Forestry Faculty and Graduate Assistants Employed in Resident Instruction, Cooperative Extension, and Research in Land Grant Institutions, Fall 1993 (full-time equivalents)

| Discipline | Faculty | Graduate Assistants |
|---|---|---|
| *Agriculture and Natural Resources* | | |
| Resident instruction | 2,537 | 613 |
| Cooperative extension (campus based) | 2,241 | 295 |
| Research | 4,693 | 1,821 |
| Agricultural experiment station | 3,913 | 1,573 |
| Other research | 780 | 248 |
| Subtotal | 9,471 | 2,729 |
| *Forestry* | | |
| Resident instruction | 228 | 61 |
| Cooperative extension (campus based) | 122 | 4 |
| Research | 538 | 388 |
| Agricultural experiment station | 334 | 242 |
| Other research | 204 | 146 |
| Subtotal | 888 | 452 |
| Total | 10,359 | 3,181 |

SOURCE: Data are from USDA Food and Agricultural Education Information System (FAEIS).

- Colleges of agriculture have developed a variety of strategies for meeting their responsibilities for research, teaching, and extension activities. A recent survey indicates that, system wide, one-half of the time of faculty in agriculture and natural resources is formally allocated to research, while the other one-half is split almost evenly between teaching and cooperative extension activities. (Of course, functional responsibilities of individual faculty do not necessarily match this aggregate norm.) The distribution of forestry faculty's time is more heavily oriented toward research (60 percent) and less oriented toward extension (only 14 percent) (Table 7-3).

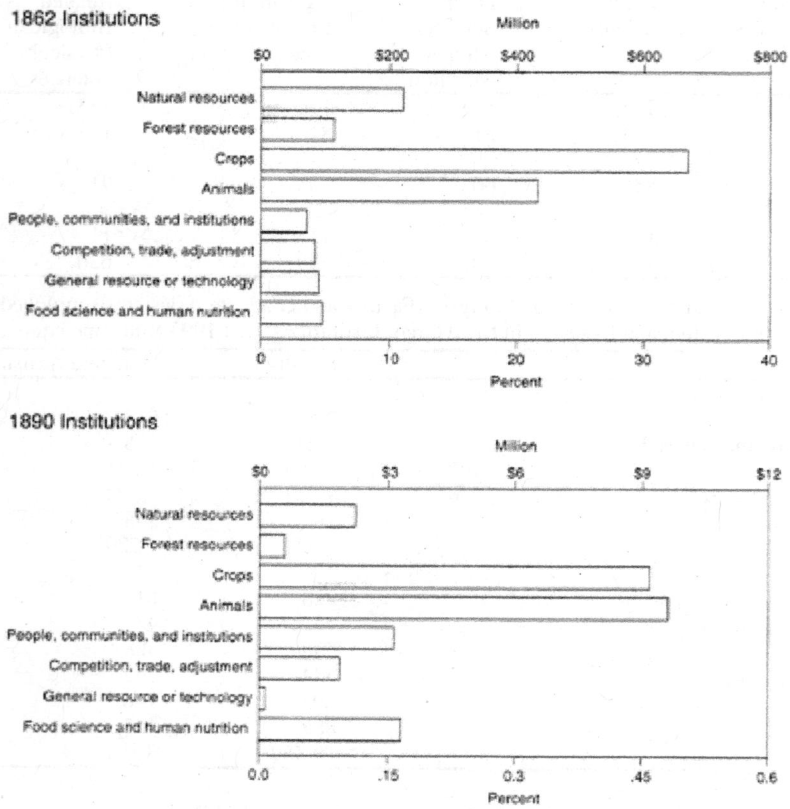


**FIGURE 7-1**

*The four graphs show the 1992 allocation of research expenditures among CRIS research program groups by 1862s (89 percent of total expenditures), 1890s (2 percent of total expenditures), forestry schools (3 percent of total expenditures), and schools of veterinary medicine (6 percent of total expenditures).*

## PROFILES-PART B SIMILARITIES AND DIFFERENCES IN RESEARCH EMPHASIS

- The CRIS reporting system asks researchers to assign each of their research projects to one of eight "research program groups" (RPGs). These RPGs include natural resources; forest resources; crops; animals; people, communities, and institutions; competition, trade, and policy; general resource or technology; and food science and human nutrition.
- The story that emerges from the allocation of research projects to these program areas is similar to that told earlier based on ESCOP program groups. System wide, research dollars are allocated first to crops, second to animals, and third to natural resources. These three program groups account for three-quarters of research expenditures by all reporting institutions, or $1.5 billion of $2 billion in total research expenditures (Figure 7-1; see also Appendix Table 4).[1]

[1] Some forestry and veterinary schools do not report to CRIS or do so partially. All 1862 SAESs and 1890 colleges must report research expenditures to USDA Current Research Information System (CRIS).

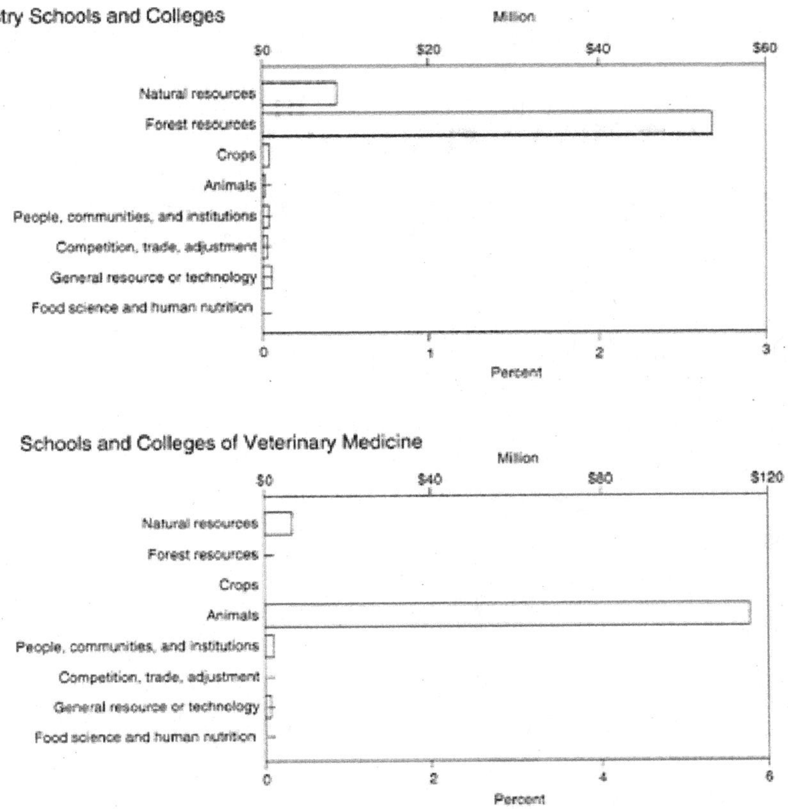


**FIGURE 7-1**

- System-wide averages may mask significant differences across individual institutions. Forestry schools are, of course, focused on forest resources research; and veterinary medicine schools are predominately conducting animals research. Among the 1862s there are some differences that make good geographic sense, but the few states that have allocations significantly different than the average are exceptions.
- Alaska's experiment station, for example, invests more in natural resources research than in animal research; Connecticut invests more in food science and human nutrition than in crops; and Vermont invests more in food and nutrition than in animals. Rhode Island puts more research money into natural resources than either crops or animals research; and West Virginia puts more into both forest and natural resources research than into crops. Cornell U. spreads research dollars more evenly than many others, but the three program groups that receive the most—crops, animals, and natural resources—are the same as the average top three (and the Geneva AES devotes most of its research to crops) (see Appendix Table 4).

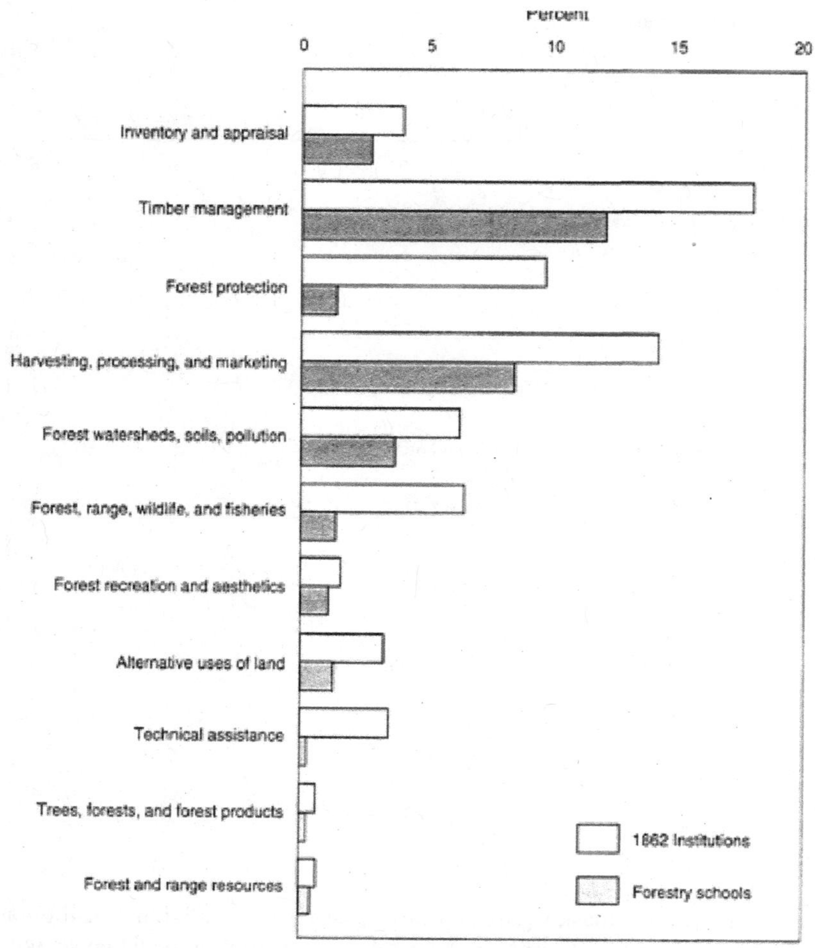


**FIGURE 7-2**

*The graph shows that most forest resources research expenditures in 1992 by 1862 institutions and forestry schools went to timber management; harvesting, processing, and marketing; and forest protection.*

- More in-depth study of individual projects is needed before it can be determined how research within a research program group differs across institutions with respect to specific focus or goal. The CRIS classification system does provide some additional breakdown. For example, when "forest resources" research is examined more closely, we can see that timber management is the primary focus and harvesting, processing, and marketing forest products is the secondary focus. These two areas of research accounted for more than 50 percent of all forest resources research in 1992 (Figure 7-2; see also Appendix Table 5).

- Some schools had a different forestry research orientation, however. For example, Vermont's forestry school focuses on forest watersheds, soils and pollution, while the universities of New Hampshire, Wyoming, and Florida, among others, devote significant shares of research dollars to the study of forest, range, wildlife, and fisheries habitat development (see Appendix Table 5).

## COLLEGES CLUSTERED ACCORDING TO COMMODITY RESEARCH

- Research on crops and livestock comprises the majority of research at 1862s, but the specific commodities studied at an institution are determined by the characteristics of the state's farm production. Figure 7-3 shows the system's allocation of commodity research expenditures in 1992 among specific commodities or commodity groupings. It also shows the percentage contribution of each commodity group to total cash receipts from farm sales.

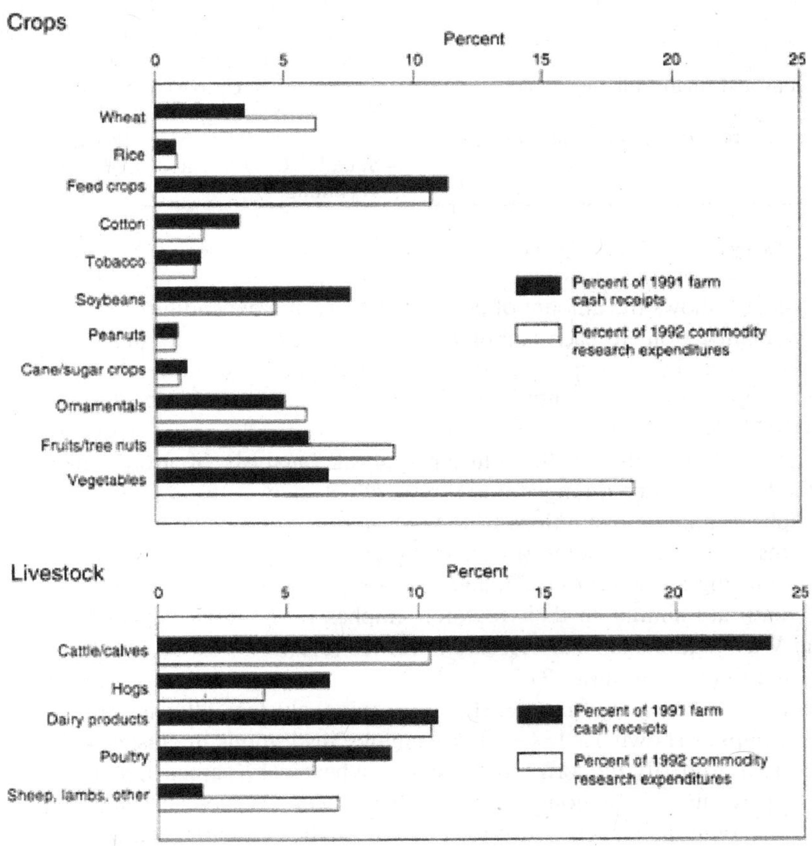


**FIGURE 7-3**

*The graphic presentation of the amounts (as percentage) of commodity research expenditures allocated to specific crops and animals research by 1862 institutions in 1992 indicates that vegetables account for a large share. Dairy and beef cattle combined, however, account for more than 20 percent.*

TABLE 7-4 Classification of 1862 State Agricultural Experiment Stations as Commodity Research Clusters

| Cluster No. | Commodity or Group | Institution |
|---|---|---|
| 1 | Beef cattle, vegetables, cotton | Auburn U., U. of Arizona, U. of Georgia, Mississippi State U., New Mexico State U., Texas A&M U. |
| 2 | Vegetables, citrus and other fruits | U. of California, U. of Florida, U. of Hawaii |
| 3 | Dairy cattle, vegetables, beef cattle | Clemson U., U. of Kentucky, U. of Maryland, North Carolina State U., U. of Tennessee |
| 4 | Corn, soybeans, swine | U. of Illinois, Iowa State U., U. of Minnesota, U. of Missouri, Purdue U. |
| 5 | Beef cattle, wheat, vegetables | Colorado State U., U. of Idaho, Kansas State U., Montana State U., U. of Nebraska, North Dakota State U., U. of Nebraska, Oklahoma State U., South Dakota State U. |
| 6 | Rice, soybeans, beef cattle | U. of Arkansas, Louisiana State U. |
| 7 | Dairy cattle, vegetables, poultry | U. of Connecticut; Cornell U.; U. of Delaware; U. of New Hampshire; Ohio State U.; Pennsylvania State U.; Rutgers-The State U. of New Jersey, Cook College; Virginia Polytechnic Inst. and State U.; U. of Wisconsin |
| 8 | Deciduous and small fruit, vegetables | Geneva AES, U. of Maine, Michigan State U., Oregon State U., Washington State U. |
| 9 | Dairy cattle, beef cattle, sheep and wool, other | U. of Alaska, U. of Massachusetts, U. of Nevada, Utah State U., U. of Vermont, West Virginia U., U of Wyoming |

NOTE: Clusters are identified by the three commodities or commodity groups allocated the greatest percentage of research funding.

- Appendix Table 6 shows the amount of commodity research expenditures for each of the 58 SAESs and the percentage allocations to each commodity for 1992. For example, in 1992 the University of Texas spent the most on commodity-specific research, and 20 percent of those expenditures went to beef cattle research. (Note: research that is not commodity specific, usually basic research applicable to multiple crops, is not included in Appendix Table 6.)
- Using a statistical procedure called cluster analysis (analysis of groups having similar patterns or profiles), the SAESs can be arranged into 9 research clusters. Table 7-4 lists the clusters by commodity research emphasis and the institutions in each cluster; Figure 7-4 shows the inclusion of each state in a commodity research cluster. Some geographic patterns emerge, such as for the "corn belt" (cluster 4), the wheat-producing states (cluster 5), and the rice producing states (cluster 6). In addition, six states across the south are similar in their research emphases on cattle, vegetables, and cotton (cluster 1). Oregon and Washington also share research profiles (cluster 8); as do North Carolina, South Carolina, Tennessee, and Kentucky (cluster 3).
- The percentage of research funds each of the nine clusters allocates to specific commodities or commodity groups is shown Table 7-5. For example, institutions in cluster 4 conduct their commodity-specific research primarily on corn (18 percent), soybeans (12 percent), and swine (15 percent), with a considerable percentage also going to research on beef (11 percent) and dairy (10 percent) cattle. Cluster 6, on the other hand, is oriented toward rice, soybeans, cattle, and poultry research; and cluster 2 focuses its research on vegetables and fruits.

For some commodities, research is concentrated at a few colleges.

| Commodity Cluster Number | Commodity or Commodity Group | Number of SAESs in Cluster |
|---|---|---|
| 1 | Beef cattle, vegetables, cotton | 6 |
| 2 | Vegetables, citrus, other fruit | 3 |
| 3 | Dairy cattle, vegetables, beef cattle | 5 |
| 4 | Corn, soybeans, swine | 5 |
| 5 | Beef cattle, wheat, vegetables | 8 |
| 6 | Rice, soybeans, beef cattle | 2 |
| 7 | Dairy cattle, vegetables, poultry | 9 |
| 8 | Deciduous, small fruit, vegetables | 5 |
| 9 | Dairy and beef cattle, sheep and wool, and other animals | 7 |

**FIGURE 7-4**

*The map shows emergent geographic patterns of commodity research at colleges of agriculture. States are identified by number indicating the profile of commodity research at their state agricultural experiment station in 1992.*

- Another way to assess the system's organization of research is to look at how expenditures for research on specific commodities are distributed among the SAESs. Table 7-6 shows that for six commodities or commodity groups—citrus and tropical fruits, cotton, peanuts, rice, sugar, and tobacco—five colleges account for more than one-half of all research expenditures. For each of these except sugar, 10 colleges perform almost all of the research.
- Crop research tends to be more concentrated than animal research; this may be because animal production is less site-specific—that is, less sensitive to climatic and geographic conditions—than crop production. Commodities for which research is least concentrated include vegetables (although research on specific vegetables may be more concentrated), poultry, dairy cattle, beef cattle, and pasture and forage crops. As a point of contrast, the five states leading dairy research account for less than 30 percent of dairy research, while the five states leading in milk production account for more than 50 percent of milk production (U.S. Department of Agriculture, 1992). Similarly, the five states leading in poultry research account for 30 percent of poultry research, while the five states leading in broiler production account for about 60 percent of the value of broiler production (U.S. Department of Agriculture, 1992).

TABLE 7-5 Commodity-Specific Research Funds (percent) as Allocated by the Nine Commodity Research Clusters

| | Commodity Research Cluster | | | | | | | | |
|---|---|---|---|---|---|---|---|---|---|
| Specific Commodity | 1 | 2 | 3 | 4 | 5 | 6 | 7 | 8 | 9 |
| Beef cattle | 15 | 7 | 10 | 11 | 25 | 12 | 5 | 4 | 16 |
| Citrus and tropical/subtropical fruit | 0 | 17 | 0 | 0 | 0 | 1 | 0 | 0 | 0 |
| Corn and grain sorghum | 6 | 2 | 5 | 18 | 6 | 3 | 6 | 1 | 1 |
| Cotton and cotton seed | 10 | 1 | 2 | 0 | 1 | 9 | 0 | 0 | 0 |
| Dairy cattle | 6 | 5 | 12 | 10 | 5 | 8 | 20 | 9 | 23 |
| Deciduous and small fruits and edible tree nuts | 6 | 10 | 6 | 2 | 2 | 5 | 8 | 29 | 9 |
| Miscellaneous and new crops | 2 | 3 | 0 | 1 | 1 | 1 | 1 | 4 | 0 |
| Vegetables and potatoes | 12 | 22 | 11 | 6 | 10 | 8 | 14 | 28 | 6 |
| Ornamentals and turf | 5 | 9 | 6 | 3 | 2 | 2 | 7 | 3 | 4 |
| Rice | 1 | 1 | 0 | 1 | 0 | 13 | 0 | 0 | 0 |
| Wheat and other small grains | 3 | 4 | 4 | 5 | 22 | 4 | 2 | 8 | 5 |
| Pasture and forage crops | 9 | 4 | 7 | 5 | 5 | 5 | 3 | 2 | 11 |
| Soybeans and other oilseed and oil crops | 6 | 1 | 7 | 12 | 7 | 13 | 4 | 1 | 0 |
| Peanuts | 4 | 1 | 1 | 0 | 1 | 0 | 0 | 0 | 0 |
| Tobacco | 1 | 0 | 8 | 0 | 0 | 0 | 1 | 0 | 0 |
| Sugar crops | 1 | 1 | 0 | 1 | 1 | 4 | 0 | 0 | 0 |
| Poultry | 7 | 3 | 9 | 5 | 1 | 11 | 14 | 3 | 5 |
| Swine | 3 | 1 | 5 | 15 | 6 | 2 | 5 | 2 | 2 |
| Sheep and wool, other animals, bees and honey, etc. | 4 | 5 | 6 | 4 | 7 | 3 | 9 | 4 | 15 |

- In Table 7-6, the five SAESs listed as the leading researchers for each commodity represent states that are either primary producers of the commodity or for which the commodity is an important agricultural product. It is perhaps not surprising that several large colleges appear often among the five leading SAESs listed for each commodity. For example, Texas A&M U. is among the leading five for 13 of 19 commodity research groups; the University of Florida appears 8 times; North Carolina State 7 times; and the University of California system 5 times.

## PROFILES-PART C COMPARISONS OF RESEARCH FUNDING MECHANISMS

- Federal funding by formula coupled with state matching grants characterizes the history of funding for each 1862 college's SAES. Today, however, the agricultural college research system encompasses a broader array of funding mechanisms. For some SAESs the private sector partners, such as commodity groups that fund research through "check-off" programs, have a growing influence over the allocation of research dollars. For others, the federal partner is increasingly Congress or a non-USDA agency.

TABLE 7-6 The Five 1862 State Agricultural Experiment Stations Conducting the Highest Percentages of Research, by Commodity or Commodity Group, 1992

| | | Percent of Research Conducted | |
|---|---|---|---|
| Commodity or Commodity Group | Five SAESs (by state)[a] Conducting the Most Research | By First SAES | By All Five SAESs |
| Beef cattle | TX, NE, FL, KS, GA | 11 | 35 |
| Citrus and tropical/subtropical fruit | FL, CA, HI, PR, TX | 40 | 96 |
| Corn and grain sorghum | IA, TX, IN, IL, NC | 11 | 43 |
| Cotton and cotton seed | TX, AZ, AR, LA, MS | 26 | 66 |
| Dairy cattle | WI, PA, NYC, TX, FL | 10 | 28 |
| Deciduous and small fruits and edible tree nuts | CA, WA, NYG, OR, NC | 15 | 42 |
| Miscellaneous and new crops | MI, WA, PR, HI, OR | 15 | 47 |
| Ornamentals and turf | FL, TX, CA, NC, GA | 13 | 34 |
| Pasture and forage crops | FL, TX, GA, MI, LA | 8 | 27 |
| Peanuts | GA, NC, TX, FL, AL | 31 | 84 |
| Poultry | NC, AR, AL, TX, GA | 7 | 30 |
| Rice | AR, LA, TX, NC, FL | 27 | 72 |
| Sheep and wool, other animals, bees and honey, etc. | CA, TX, KY, OR, NE | 10 | 34 |
| Soybeans and other oilseed and oil crops | IA, AR, IL, LA, GA | 10 | 34 |
| Sugar Crops | LA, PR, TX, FL, ID | 20 | 66 |
| Swine | IA, MN, IL, NE, IN | 14 | 42 |
| Tobacco | NC, KY, SC, TN, CTH | 42 | 74 |
| Vegetables and potatoes | FL, CA, NYC, WA, NC | 10 | 32 |
| Wheat and other small grains | KS, ND, WA, TX, OR | 13 | 40 |

Abbreviations: NYC, Cornell University; NYG, Geneva Agricultural Experiment Station; CTH, New Haven Agricultural Experiment Station. All other institutions are identified by the U.S. postal code state abbreviations.

[a] The five SAESs are ordered based on which conducts the highest percentage of research for the commodity or commodity group, the second highest, etc.

SOURCE: Data are from USDA Current Research Information System (CRIS).

- Table 7-7 ranks SAESs at 1862 colleges of agriculture by amount of their total research expenditures and provides a breakdown of funding sources for 1992 (see also Figure 7-5). Colleges with the largest research programs tend to be the recipients of the largest amount of formula funds. But it is not so much the amount received in formula-based grants that sets them apart from the smaller colleges in the system as it is the amounts they receive in other types of funding. For example, the amount of formula funds received by Texas A&M U. is only 3 times greater than the amount received by the University of Massachusetts, however, "other federal funds" are 26 times greater, state appropriations are 15 times greater, and private funds are 4 times greater.
- Thus although "big" colleges of agriculture (measured by research expenditures) get the most in formula funds, they rely on them the least. Texas A&M U. draws on formula funds for less than 7 percent of its research expenditures, while the University of Massachusetts counts on them to fund 19 percent of research. Schools that do the most research (in absolute terms) also tend to have generally more diversified funding portfolios, although there are exceptions.

TABLE 7-7 Total Research Expenditures (thousands of dollars) of State Agricultural Experiment Stations, by Institution and Funding Mechanism, 1992

| Institution | Funding Mechanism | | | | | | |
|---|---|---|---|---|---|---|---|
| | Formula Funds[a] | Competitive Research Grants[a] | Special Research Grants[a] | Other Federal Funds | State Appropriations | Private Funds | Total Expenditures |
| U. of California | 5,348 (3.6) | 8,604 (5.8) | 3,905 (2.6) | 20,928 (14.1) | 93,530 (63.1) | 15,899 (10.7) | 148,214 |
| Texas A&M U. | 6,379 (6.8) | 2,860 (3.0) | 1,835 (2.0) | 13,973 (14.9) | 50,138 (53.4) | 18,724 (19.9) | 93,910 |
| U. of Florida | 3,272 (3.7) | 1,549 (1.7) | 2,036 (2.3) | 9,235 (10.4) | 59,748 (67.1) | 13,241 (14.9) | 89,081 |
| U. of Wisconsin | 5,223 (7.9) | 4,024 (6.1) | 770 (1.2) | 19,361 (29.2) | 26,183 (39.5) | 10,650 (16.1) | 66,210 |
| Cornell U. | 4,266 (6.6) | 2,761 (4.3) | 1,877 (2.9) | 13,016 (20.1) | 19,789 (30.6) | 22,953 (35.5) | 64,662 |
| North Carolina State U. | 6,042 (9.6) | 1,623 (2.6) | 708 (1.1) | 9,541 (15.1) | 34,205 (54.3) | 10,925 (17.3) | 63,044 |
| Michigan State U. | 4,825 (8.0) | 2,148 (3.6) | 6,573 (10.9) | 13,335 (22.1) | 22,962 (38.1) | 10,462 (17.3) | 60,304 |
| Iowa State U. | 5,613 (9.4) | 1,406 (2.4) | 6,485 (10.9) | 9,329 (15.7) | 21,957 (36.8) | 14,799 (24.8) | 59,588 |
| U. of Minnesota | 5,062 (8.7) | 1,325 (2.3) | 1,530 (2.6) | 3,959 (6.8) | 32,519 (55.9) | 13,826 (23.7) | 58,221 |
| U. of Nebraska | 3,347 (6.5) | 1,055 (2.0) | 2,914 (5.6) | 5,035 (9.7) | 23,311 (45.0) | 16,195 (31.2) | 51,858 |
| Purdue U. | 4,939 (9.6) | 1,670 (3.3) | 2,887 (5.6) | 8,150 (15.9) | 21,557 (42.0) | 12,079 (23.6) | 51,281 |
| U. of Georgia | 4,949 (9.9) | 695 (1.4) | 1,768 (3.5) | 2,797 (5.6) | 35,941 (71.7) | 3,969 (7.9) | 50,119 |
| Virginia Polytechnic Institute and State U. | 4,383 (10.1) | 833 (1.9) | 358 (0.8) | 7,325 (16.8) | 22,183 (51.0) | 8,448 (19.4) | 43,530 |
| Oregon State U. | 2,661 (6.3) | 1,951 (4.6) | 2,774 (6.6) | 8,953 (21.3) | 18,047 (42.9) | 7,699 (18.3) | 42,085 |
| Kansas State U. | 3,451 (8.3) | 1,274 (3.1) | 1,383 (3.3) | 4,853 (11.7) | 20,854 (50.2) | 9,725 (23.4) | 41,540 |
| U. of Arizona | 1,955 (4.9) | 2,022 (5.1) | 378 (1.0) | 9,864 (24.8) | 19,943 (50.2) | 5,569 (14.0) | 39,731 |
| Louisiana State U. | 3,332 (8.8) | 162 (0.4) | 2,587 (6.8) | 992 (2.6) | 22,769 (59.9) | 8,173 (21.5) | 38,014 |
| U. of Illinois | 5,303 (14.1) | 1,877 (5.0) | 733 (1.9) | 3,365 (8.9) | 14,114 (37.4) | 12,319 (32.7) | 37,711 |
| Ohio State U. | 5,678 (15.1) | 685 (1.8) | 1,411 (3.7) | 2,168 (5.8) | 20,885 (55.5) | 6,809 (18.1) | 37,636 |
| Pennsylvania State U. | 6,091 (16.6) | 566 (1.5) | 2,140 (5.8) | 2,891 (7.9) | 19,111 (52.0) | 5,968 (16.2) | 36,767 |
| Washington State U. | 3,759 (10.2) | 640 (1.7) | 2,527 (6.9) | 3,916 (10.7) | 17,284 (47.1) | 8,608 (23.4) | 36,735 |
| Mississippi State U. | 4,368 (12.3) | 280 (0.8) | 3,543 (10.0) | 2,801 (7.9) | 16,165 (45.6) | 8,304 (23.4) | 35,462 |
| U. of Missouri | 4,778 (14.1) | 2,193 (6.5) | 1,699 (5.0) | 2,055 (6.0) | 16,876 (49.6) | 6,393 (18.8) | 33,993 |
| Auburn U. | 4,268 (13.2) | 792 (2.4) | 799 (2.5) | 1,840 (5.7) | 17,183 (53.0) | 7,568 (23.3) | 32,451 |
| U. of Arkansas | 3,728 (11.5) | 506 (1.6) | 2,789 (8.6) | 1,791 (5.5) | 18,124 (56.1) | 5,377 (16.6) | 32,315 |
| Colorado State U. | 2,728 (9.1) | 1,170 (3.9) | 2,197 (7.3) | 10,788 (36.1) | 7,709 (25.8) | 5,299 (17.7) | 29,892 |
| Clemson U. | 3,138 (11.0) | 478 (1.7) | 837 (2.9) | 1,617 (5.6) | 19,085 (66.6) | 3,489 (12.2) | 28,644 |
| Oklahoma State U. | 3,247 (11.5) | 866 (3.1) | 862 (3.1) | 1,683 (6.0) | 16,276 (57.7) | 5,258 (18.7) | 28,192 |
| North Dakota State U. | 2,308 (8.8) | 472 (1.8) | 2,511 (9.5) | 2,296 (8.7) | 12,981 (49.3) | 5,766 (21.9) | 26,334 |
| U. of Kentucky | 4,997 (19.5) | 651 (2.5) | 597 (2.3) | 0 (0.0) | 18,374 (71.5) | 1,064 (4.1) | 25,682 |
| U. of Tennessee | 4,861 (19.3) | 994 (4.0) | 138 (0.5) | 836 (3.3) | 14,013 (55.7) | 4,295 (17.1) | 25,137 |

| | Funding Mechanism | | | | | | |
|---|---|---|---|---|---|---|---|
| Institution | Formula Funds[a] | Competitive Research Grants[a] | Special Research Grants[a] | Other Federal Funds | State Appropriations | Private Funds | Total Expenditures |
| Rutgers—The State U., Cook College | 2,762 (11.0) | 1,502 (6.0) | 1,324 (5.3) | 2,475 (9.9) | 12,985 (51.8) | 3,998 (16.0) | 25,046 |
| U. of Hawaii | 1,309 (5.6) | 311 (1.3) | 3,061 (13.2) | 2,881 (12.4) | 12,897 (55.5) | 2,765 (11.9) | 23,224 |
| U. of Maryland | 2,504 (12.0) | 666 (3.2) | 568 (2.7) | 3,058 (14.6) | 12,797 (61.3) | 1,282 (6.1) | 20,875 |
| U. of Idaho | 1,965 (10.8) | 495 (2.7) | 1,000 (5.5) | 2,117 (11.6) | 10,258 (56.4) | 2,356 (13.0) | 18,191 |
| Utah State U. | 1,818 (10.7) | 613 (3.6) | 138 (0.8) | 3,736 (22.0) | 7,294 (42.9) | 3,402 (20.0) | 17,002 |
| Montana State U. | 1,980 (12.4) | 288 (1.8) | 468 (2.9) | 2,754 (17.2) | 7,502 (47.0) | 2,978 (18.6) | 15,971 |
| South Dakota State U. | 2,452 (16.9) | 657 (4.5) | 82 (0.6) | 355 (2.4) | 6,457 (44.6) | 4,477 (30.9) | 14,480 |
| New Mexico State U. | 1,724 (13.3) | 460 (3.5) | 1,243 (9.6) | 1,350 (10.4) | 7,414 (57.1) | 790 (6.1) | 12,981 |
| U. of Maine | 2,233 (18.1) | 378 (3.1) | 673 (5.5) | 1,281 (10.4) | 5,715 (46.3) | 2,050 (16.6) | 12,331 |
| U. of Massachusetts | 2,259 (19.5) | 429 (3.7) | 485 (4.2) | 522 (4.5) | 3,180 (27.4) | 4,727 (40.7) | 11,602 |
| U. of Puerto Rico | 3,872 (34.1) | 0 (0.0) | 332 (2.9) | 0 (0.0) | 6,397 (56.3) | 752 (6.6) | 11,353 |
| Geneva AES | 864 (7.8) | 284 (2.5) | 659 (5.9) | 726 (6.5) | 6,269 (56.3) | 2,341 (21.0) | 11,143 |
| U. of Delaware | 1,299 (13.8) | 250 (2.6) | 5 (0.1) | 408 (4.3) | 4,968 (52.6) | 2,516 (26.6) | 9,445 |
| West Virginia U. | 2,824 (34.3) | 198 (2.4) | 45 (0.5) | 1,015 (12.3) | 3,258 (39.6) | 885 (10.8) | 8,226 |
| U. of Nevada | 1,192 (15.8) | 0 (0.0) | 5 (0.1) | 1,033 (13.7) | 4,014 (53.3) | 1,282 (17.0) | 7,526 |
| U. of Vermont | 1,322 (18.4) | 42 (0.6) | 1,814 (25.2) | 1,267 (17.6) | 1,924 (26.8) | 819 (11.4) | 7,187 |
| U. of Wyoming | 1,670 (26.7) | 361 (5.8) | 29 (0.5) | 340 (5.4) | 3,591 (57.4) | 269 (4.3) | 6,260 |
| U. of Connecticut | 1,005 (16.7) | 233 (3.9) | 378 (6.3) | 934 (15.5) | 2,312 (38.5) | 1,151 (19.1) | 6,011 |
| U. of Alaska | 1,269 (22.3) | 19 (0.3) | 5 (0.1) | 512 (9.0) | 3,473 (61.1) | 407 (7.2) | 5,685 |
| New Haven AES | 872 (16.9) | 172 (3.3) | 114 (2.2) | 192 (3.7) | 3,743 (72.4) | 76 (1.5) | 5,170 |
| U. of New Hampshire | 1,619 (34.6) | 204 (4.4) | 101 (2.2) | 0 (0.0) | 2,468 (52.7) | 289 (6.2) | 4,681 |
| U. of Rhode Island | 1,237 (34.6) | 50 (1.4) | 166 (4.6) | 390 (10.9) | 1,663 (46.5) | 71 (2.0) | 3,577 |
| U. of Guam | 781 (28.7) | 0 (0.0) | 279 (10.2) | 0 (0.0) | 1,664 (61.1) | 0 (0.0) | 2,724 |
| U. of the Virgin Islands | 735 (52.0) | 0 (0.0) | 214 (15.1) | 0 (0.0) | 464 (32.8) | 0 (0.0) | 1,413 |
| U. of the District of Columbia | 498 (61.8) | 0 (0.0) | 0 (0.0) | 0 (0.0) | 307 (38.2) | 0 (0.0) | 805 |
| American Samoa Community College | 641 (100.0) | 0 (0.0) | 0 (0.0) | 0 (0.0) | 0 (0.0) | 0 (0.0) | 641 |
| Northern Marianas College | 454 (79.1) | 0 (0.0) | 0 (0.0) | 0 (0.0) | 0 (0.0) | 120 (20.9) | 575 |
| Total | 177,459 (18.0) | 55,745 (2.9) | 76,742 (4.8) | 226,037 (10.2) | 906,830 (48.5) | 329,655 (16.2) | 1,772,467 |

NOTE: Some SAESs may have incompletely reported research expenditures to CRIS, which would affect the accuracy of this table. For example, in 1992 University of Kentucky's SAES received $725,000 from "other federal funds," which was not reported to CRIS because of university reporting procedures in that year (James A. Boling, University of Kentucky, personal communication, 1995). Number in parentheses is percent of total research expenditures.

[a] Grant programs administered by USDA Cooperative State Research, Education, and Extension Service.

SOURCE: Data are from USDA Current Research Information System (CRIS).

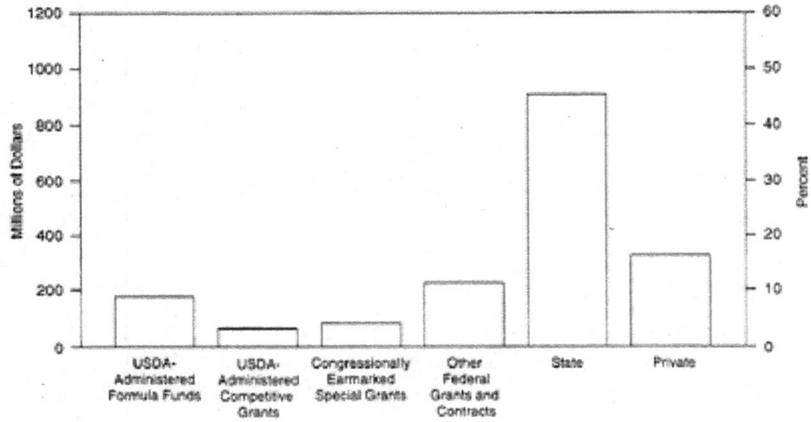


**FIGURE 7-5**

*In 1992 more than 40 percent of total research expenditures by state agricultural experiment stations was provided by the states.*

- Even colleges that have research programs similar in size have taken different funding paths. For example, the University of Missouri and Mississippi State U. had similar research expenditures in 1992; their levels of support from formula funds, other federal funds, and state appropriations were extremely close. However, Mississippi had more private support and nearly twice as much as Missouri in congressionally earmarked grants; Missouri, on the other hand, received significantly more in USDA competitive grants than did Mississippi.

## COLLEGES CLUSTERED ACCORDING TO RESEARCH FUNDING MECHANISMS

- Cluster analysis can also be used to group SAESs according to their funding profile. Funding profiles derived here are based on an institution's portfolio of nonstate funding sources—formula funds, USDA competitive grants, special grants, other federal grants, and private funds—and the percentage of total nonstate funding each source represents in just a single year—1992. Clusters, then, are determined on the basis of the prominence of a particular funding mechanism, such as formula funding, in that year. Table 7-8 classifies 1862 SAESs according to five research funding clusters. Table 7-9 provides a breakdown, in percent, of federal and private funding sources to each of the five funding clusters.
- Colleges in cluster 1 depended, on average, on traditional formula funds for only 17 percent of their nonstate funds. They received significant funding from other federal funds and private funds. The major land grant research universities are in this group. The funding portfolio for cluster 2 (the largest group) is also quite diversified—on average, formula funds accounted for only 26 percent of nonstate funds—but cluster 2 received more from private funds and less from other federal funds than cluster 1. In other words, cluster 2 colleges appear to be slightly less oriented toward participation in the competitive grant programs of federal agencies and slightly more oriented toward commercial sales of products and licenses and private-sector partnerships.
- In 1992 colleges in cluster 3 received, on average, more than 40 percent of their nonstate funding from formula grants but also received more than 25 percent of their nonstate funds in the form of special grants—that is, research grants earmarked by Congress. Clusters 4 and 5 rely more on formula funds, although colleges in cluster 4 received relatively more funding through competitive grants than did those in cluster 5.

TABLE 7-8 Classification of 1862 State Agricultural Experiment Stations as Research Funding Clusters, 1992

| Cluster No. | Primary Funding Mechanism | SAES Affiliate Institution or Location |
|---|---|---|
| 1 | Other federal (36%), private (35%), formula (17%) | U. or Arizona, U. of California agricultural research system, Colorado State U., Cornell U., U. of Florida, Kansas State U., U. of Maryland, Michigan State U., Montana State U., U. of Nevada, North Carolina State U., Oregon State U., Purdue U., Texas A&M U., Utah State U., Virginia Polytechnic Institute and State U., U. of Wisconsin |
| 2 | Private (43%), formula (26%) | Auburn U.; U. of Arkansas; Clemson U.; U. of Connecticut; U. of Delaware; Geneva AES; U. of Georgia; U. of Idaho; U. of Illinois; Iowa State U.; Louisiana State U.; U. of Maine; U. of Massachusetts; U. of Minnesota; Mississippi State U.; U. of Missouri; U. of Nebraska; North Dakota State U.; Ohio State U.; Oklahoma State U.; Pennsylvania State U.; Rutgers-The State U., Cook College; South Dakota State U.; U. of Tennessee; Washington State U. |
| 3 | Formula (44%), special (27%) | U. of Guam, U. of Hawaii, New Mexico State U., U. of Vermont, U. of the Virgin Islands |
| 4 | Formula (66%) | U. of Kentucky, U. of New Hampshire, New Haven AES, U. of Wyoming |
| 5 | Formula (77%) | American Samoa Community College, U. of Alaska, U. of the District of Columbia, Northern Marianas College, U. of Puerto Rico, U. of Rhode Island, West Virginia U. |

NOTE: Some SAESs may have incompletely reported research expenditures to CRIS, which would affect the accuracy of the membership of each research funding cluster. For example, in 1992 University of Kentucky's SAES received $725,000 from "other federal funds," which was not reported to CRIS because of university reporting procedures in that year (James A. Boling, University of Kentucky, personal communication, 1995). This and other potential reporting omissions may affect the classification of the SAESs by research funding cluster. Funding does not include state funding.

TABLE 7-9 Breakdown (percent) of Total Federal and Private Research Funding to 1862 State Agricultural Experiment Funding Clusters

| | Percent Provided by Funding Mechanisms | | | | | |
|---|---|---|---|---|---|---|
| Cluster No. | Formula Funding[a] | Competitive Grants[a] | USDA Special Grants[a] | Other Federal Grants | Private Funds | Total |
| 1 | 17 | 7 | 6 | 36 | 35 | 101 |
| 2 | 26 | 6 | 10 | 15 | 43 | 100 |
| 3 | 44 | 2 | 27 | 15 | 11 | 99 |
| 4 | 66 | 11 | 6 | 7 | 11 | 101 |
| 5 | 77 | 1 | 2 | 9 | 11 | 100 |

NOTE: Totals may not equal 100 because of rounding.

[a] Grant programs administered by USDA Cooperative State Research, Education, and Extension Service.

SOURCE: Data are from USDA Current Information Research System (CRIS).

- In 1991 28 land grant universities were among the leading 100 university recipients of federal funds for research and development. These are ranked, along with each university's three primary federal funding sources in Table 7-10. The table indicates how far from their agriculture roots some of the land grant universities have come. For only 4 of the 28—North Carolina State U., Iowa State U., Virginia Polytechnic Institute and State U., and the University of Nebraska—is USDA the primary federal funding agency. However, at these schools, USDA funding still accounts for more than 20 percent of federal research and development funds.

TABLE 7-10 Amount (percent) of Funds Received for Research and Development by 1862 Land Grant Universities from Their Three Primary Federal Funding Agencies, 1991

| | Percent of Total from the Three Primary Contributing Agencies | | | | | | |
|---|---|---|---|---|---|---|---|
| Institution | HHS | NSF | DOE | DOD | NASA | USDA | AOAs |
| U. of Wisconsin, Madison | 52.7 | 18.4 | 9.6 | | | | |
| U. of Minnesota | 61.4 | 14.0 | | | | | 7.4 |
| Cornell U. | 44.0 | 29.9 | | 10.8 | | | |
| Pennsylvania State U. | 24.4 | 12.4 | | 40.7 | | | |
| U. of California, Berkeley | 33.9 | 25.1 | | | 14.4 | | |
| U. of Illinois, Urbana-Champaign | 17.0 | 43.3 | | 17.5 | | | |
| U. of Arizona | 53.0 | 14.4 | | | 14.9 | | |
| Ohio State U. | 43.0 | 19.3 | | | | 11.0 | |
| U. of California, Davis | 55.3 | 12.1 | 15.3 | | | | |
| U. of Massachusetts | 52.7 | 17.1 | | 12.6 | | | |
| U. of Maryland, College Park | | 25.2 | | 20.1 | 23.1 | | |
| Louisiana State U. System | 38.1 | | 36.1 | | | 10.3 | |
| Purdue U. | 30.2 | 27.4 | | | | 14.9 | |
| U. of Florida | 52.6 | 11.3 | | | | 12.2 | |
| Michigan State U. | 29.6 | 33.0 | | | | 31.7 | |
| Texas A&M U. | 20.6 | 27.2 | | | | 20.6 | |
| Oregon State U. | | 26.7 | | | | 19.9 | 19.2 |
| U. of Tennessee (all campuses) | 46.8 | | 23.1 | | | 11.3 | |
| Rutgers-The State U. | 44.3 | 27.9 | | | | 10.0 | |
| North Carolina State U., Raleigh | | 20.9 | | | | 23.1 | 17.4 |
| Colorado State U. | 39.4 | 19.6 | | | | 14.0 | |
| U. of Hawaii, Manoa | 25.7 | 32.7 | | | 14.9 | | |
| U. of Vermont and State Agriculture College | 75.9 | 6.0 | | | | 1.5 | |
| U. of Kentucky | 54.2 | 11.1 | | | | 19.5 | |
| Iowa State U. | 18.7 | | | | | 38.0 | 20.2 |
| Virginia Polytechnic Institute and State U. | | 15.7 | 17.2 | | | 20.1 | |
| U. of Nebraska, Lincoln | | 23.2 | 22.2 | | | 26.1 | |
| U. of Missouri, Columbia | 50.9 | 11.4 | | | | 29.2 | |

NOTE: These are the land grant universities included in the top 100 universities receiving federal funds. They are ranked here according to the total amount of funds received. Abbreviations used: HHS, U.S. Department of Health and Human Services; NSF, National Science Foundation; DOE, Department of Energy; DOD, Department of Defense; NASA, National Aeronautics and Space Administration; USDA, U.S. Department of Agriculture; AOAs, all other agencies.

SOURCE: Adapted from National Science Foundation. 1991. Federal Support to Universities, Colleges, and Nonprofit Institutions: Fiscal Year 1991. Arlington, Va.: National Science Foundation.

- The university-wide research environment may be very important to the research funding prospects of the university's college of agriculture. Note that 13 of the 17 colleges in cluster 1 in Table 7-7—the cluster least reliant on USDA formula funding and most diversified toward a combination of other federal funds and USDA competitive grants—are found at land grant universities that receive large percentages of federal research and development funding.

## ISSUES FOR DISCUSSION

- Colleges of agriculture are changing their names at a rapid pace. To what extent do the name changes reflect significant changes in college programs? To what extent does the growing diversity of college names reflect real differences among individual colleges?
- Do differences in colleges' administrative structure, and how these structures interface with the university-wide administration, make a difference to college performance?
- Most colleges direct a majority of their research expenditures to plant and animal systems research, but the specific crops and animals of most interest vary regionally. Often, several colleges in the same geographic region share similar commodity research profiles. How does the system avoid redundancy of research effort? Are there opportunities for additional regional research collaboration? What are the advantages and disadvantages of the "regionalization" of research?
- Some commodity research is highly concentrated at a small number of colleges, but for a number of commodities research is still significantly more diffused than production of the same commodity. Does formula funding provide continued support for smaller, state-based commodity research programs that might not otherwise survive? What would be the benefits and costs of greater specialization in commodity-specific research within the college system?
- Traditional formula funding has decreased in importance to the large research colleges in the system, but it is still considerably important to many of the smaller ones. How would changes in the formula, or changes in the overall percentage of support through formula funds, affect the distribution of funds among colleges and the viability of smaller institutions?

# APPENDIX

The compilation of data for this publication generated a great deal of information in extensive detail. The tables in this appendix expand on data presented in the text and provide a more detailed picture of the land grant colleges and universities and the constituency they serve. Table 1 provides, for each state, a breakdown of the distribution of U.S. population as urban, suburban, rural farm, and rural nonfarm. Table 2 lists, for each region, the total number of graduates in agriculture, food, and natural resources disciplines for all degree programs for land grant and non-land grant schools. Table 3 provides, in detail, the allocation of extension staff (number and percent) for the seven base program groups by each of the states. Table 4 provides the allocation of research expenditures (amount and percent) for the eight CRIS research program groups by all 1862s, 1890s, and related forestry schools and schools of veterinary medicine. Tables 5 and 6 provide information in similar detail for forest resources research expenditures and commodity-specific research expenditures, respectively.

APPENDIX TABLE 1 Distribution of U.S. Population, 1990

| State | Urban Total (thousands) | Urban % U.S. Urban | Urban % State | Suburban Total (thousands) | Suburban % U.S. Suburban | Suburban % State |
|---|---|---|---|---|---|---|
| Alabama | 1,840 | 1.1 | 45.5 | 598 | 2.0 | 14.8 |
| Alaska | 222 | 0.1 | 40.3 | 149 | 0.5 | 27.1 |
| Arizona | 2,656 | 1.7 | 72.5 | 551 | 1.8 | 15.0 |
| Arkansas | 592 | 0.4 | 25.2 | 667 | 2.2 | 28.4 |
| California | 25,466 | 15.8 | 85.6 | 2,106 | 7.1 | 7.1 |
| Colorado | 2,378 | 1.5 | 72.2 | 338 | 1.1 | 10.3 |
| Connecticut | 2,456 | 1.5 | 74.7 | 146 | 0.5 | 4.4 |
| Delaware | 459 | 0.3 | 69.0 | 28 | 0.1 | 4.2 |
| District of Columbia | 607 | 0.4 | 100.0 | 0 | 0 | 0 |
| Florida | 10,181 | 6.3 | 78.7 | 789 | 2.6 | 6.1 |
| Georgia | 3,260 | 2.0 | 50.3 | 836 | 2.8 | 12.9 |
| Hawaii | 747 | 0.5 | 67.4 | 239 | 0.8 | 21.5 |
| Idaho | 278 | 0.2 | 27.7 | 300 | 1.0 | 29.8 |
| Illinois | 8,479 | 5.3 | 74.2 | 1,190 | 4.0 | 10.4 |
| Indiana | 2,691 | 1.7 | 48.5 | 905 | 3.0 | 16.3 |
| Iowa | 942 | 0.6 | 33.9 | 740 | 2.5 | 26.7 |
| Kansas | 1,019 | 0.6 | 41.1 | 694 | 2.3 | 28.0 |
| Kentucky | 1,277 | 0.8 | 34.6 | 633 | 2.1 | 17.2 |
| Louisiana | 2,228 | 1.4 | 52.8 | 644 | 2.2 | 15.3 |
| Maine | 267 | 0.2 | 21.7 | 282 | 0.9 | 22.9 |
| Maryland | 3,581 | 2.2 | 74.9 | 307 | 1.0 | 6.4 |
| Massachusetts | 4,730 | 2.9 | 78.6 | 339 | 1.1 | 5.6 |
| Michigan | 5,812 | 3.6 | 62.5 | 743 | 2.5 | 8.0 |
| Minnesota | 2,370 | 1.5 | 54.2 | 685 | 2.3 | 15.7 |
| Mississippi | 618 | 0.4 | 24.0 | 594 | 2.0 | 23.1 |
| Missouri | 2,783 | 1.7 | 54.4 | 733 | 2.5 | 14.3 |
| Montana | 209 | 0.1 | 26.1 | 211 | 0.7 | 26.4 |
| North Carolina | 2,511 | 1.6 | 37.9 | 825 | 2.8 | 12.4 |
| North Dakota | 202 | 0.1 | 31.7 | 138 | 0.5 | 21.6 |
| Nebraska | 688 | 0.4 | 43.6 | 356 | 1.2 | 22.6 |
| Nevada | 911 | 0.6 | 75.8 | 150 | 0.5 | 12.5 |
| New Hampshire | 340 | 0.2 | 30.6 | 226 | 0.8 | 20.4 |
| New Jersey | 6,630 | 4.1 | 85.8 | 280 | 0.9 | 3.6 |
| New Mexico | 649 | 0.4 | 42.9 | 455 | 1.5 | 30.1 |
| New York | 14,117 | 8.8 | 78.5 | 1,048 | 3.5 | 5.8 |
| Ohio | 6,656 | 4.1 | 61.4 | 1,383 | 4.6 | 12.8 |
| Oklahoma | 1,355 | 0.8 | 43.1 | 775 | 2.6 | 24.7 |
| Oregon | 1,420 | 0.9 | 50.0 | 583 | 2.0 | 20.5 |
| Pennsylvania | 7,210 | 4.5 | 60.7 | 981 | 3.3 | 8.3 |
| Puerto Rico | 2,126 | 1.3 | 60.4 | 383 | 1.3 | 10.9 |
| Rhode Island | 825 | 0.5 | 82.2 | 39 | 0.1 | 3.9 |
| South Carolina | 1,425 | 0.9 | 40.9 | 480 | 1.6 | 13.8 |
| South Dakota | 164 | 0.1 | 23.6 | 184 | 0.6 | 26.4 |
| Tennessee | 2,217 | 1.4 | 45.5 | 752 | 2.5 | 15.4 |
| Texas | 11,374 | 7.1 | 67.0 | 2,263 | 7.6 | 13.3 |
| Utah | 1,320 | 0.8 | 76.6 | 180 | 0.6 | 10.4 |
| Vermont | 87 | 0.1 | 15.4 | 94 | 0.3 | 16.7 |
| Virginia | 3,830 | 2.4 | 61.9 | 464 | 1.6 | 7.5 |
| Washington | 3,214 | 2.0 | 66.0 | 503 | 1.7 | 10.3 |
| West Virginia | 389 | 0.2 | 21.7 | 259 | 0.9 | 14.4 |
| Wisconsin | 2,465 | 1.5 | 50.4 | 747 | 2.5 | 15.3 |
| Wyoming | 114 | 0.1 | 25.1 | 180 | 0.6 | 39.8 |
| Total | 160,384 | 100.0 | 63.6 | 29,177 | 100.0 | 11.6 |

SOURCE: Data are from the U.S. Bureau of the Census.

| State | Rural Farm Total (thousands) | % U.S. Farm | % State | Rural Nonfarm Total (thousands) | % U.S. Nonfarm | % State | State Total (thousands) | % U.S. Total |
|---|---|---|---|---|---|---|---|---|
| Alabama | 59 | 1.5 | 1.5 | 1,544 | 2.6 | 38.2 | 4,041 | 1.6 |
| Alaska | 1 | <0.1 | 0.2 | 178 | 0.3 | 32.4 | 550 | 0.2 |
| Arizona | 7 | 0.2 | 0.2 | 451 | 0.8 | 12.3 | 3,665 | 1.4 |
| Arkansas | 64 | 1.6 | 2.7 | 1,029 | 1.7 | 43.8 | 2,351 | 0.9 |
| California | 151 | 3.8 | 0.5 | 2,038 | 3.4 | 6.8 | 29,760 | 11.7 |
| Colorado | 45 | 1.1 | 1.4 | 534 | 0.9 | 16.2 | 3,294 | 1.3 |
| Connecticut | 5 | 0.1 | 0.2 | 680 | 1.1 | 20.7 | 3,287 | 1.3 |
| Delaware | 6 | 0.2 | 1.0 | 172 | 0.3 | 25.9 | 666 | 0.3 |
| District of Columbia | 0 | 0 | 0 | 0 | 0 | 0 | 607 | 0.2 |
| Florida | 47 | 1.2 | 0.4 | 1,920 | 3.2 | 14.8 | 12,938 | 5.1 |
| Georgia | 80 | 2.0 | 1.2 | 2,302 | 3.8 | 35.5 | 6,478 | 2.5 |
| Hawaii | 6 | 0.2 | 0.6 | 116 | 0.2 | 10.5 | 1,108 | 0.4 |
| Idaho | 45 | 1.1 | 4.5 | 384 | 0.6 | 38.1 | 1,007 | 0.4 |
| Illinois | 207 | 5.2 | 1.8 | 1,555 | 2.6 | 13.6 | 11,431 | 4.5 |
| Indiana | 188 | 4.8 | 3.4 | 1,760 | 2.9 | 31.7 | 5,544 | 2.2 |
| Iowa | 257 | 6.5 | 9.2 | 837 | 1.4 | 30.2 | 2,777 | 1.1 |
| Kansas | 108 | 2.7 | 4.4 | 657 | 1.1 | 26.5 | 2,478 | 1.0 |
| Kentucky | 174 | 4.4 | 4.7 | 1,601 | 2.7 | 43.4 | 3,685 | 1.4 |
| Louisiana | 40 | 1.0 | 1.0 | 1,308 | 2.2 | 31.0 | 4,220 | 1.7 |
| Maine | 11 | 0.3 | 0.9 | 669 | 1.1 | 54.5 | 1,228 | 0.5 |
| Maryland | 33 | 0.8 | 0.7 | 861 | 1.4 | 18.0 | 4,781 | 1.9 |
| Massachusetts | 9 | 0.2 | 0.2 | 938 | 1.6 | 15.6 | 6,016 | 2.4 |
| Michigan | 120 | 3.0 | 1.3 | 2,620 | 4.4 | 28.2 | 9,295 | 3.7 |
| Minnesota | 208 | 5.3 | 4.8 | 1,111 | 1.9 | 25.4 | 4,375 | 1.7 |
| Mississippi | 56 | 1.4 | 2.2 | 1,306 | 2.2 | 50.7 | 2,573 | 1.0 |
| Missouri | 180 | 4.6 | 3.5 | 1,421 | 2.4 | 27.8 | 5,117 | 2.0 |
| Montana | 46 | 1.2 | 5.7 | 333 | 0.6 | 41.7 | 799 | 0.3 |
| North Carolina | 117 | 3.0 | 1.8 | 3,176 | 5.3 | 47.9 | 6,629 | 2.6 |
| North Dakota | 60 | 1.5 | 9.4 | 238 | 0.4 | 37.3 | 639 | 0.3 |
| Nebraska | 118 | 3.0 | 7.5 | 417 | 0.7 | 26.4 | 1,578 | 0.6 |
| Nevada | 5 | 0.1 | 0.4 | 136 | 0.2 | 11.3 | 1,202 | 0.5 |
| New Hampshire | 6 | 0.1 | 0.5 | 538 | 0.9 | 48.5 | 1,109 | 0.4 |
| New Jersey | 17 | 0.4 | 0.2 | 802 | 1.3 | 10.4 | 7,730 | 3.0 |
| New Mexico | 15 | 0.4 | 1.0 | 395 | 0.7 | 26.1 | 1,515 | 0.6 |
| New York | 82 | 2.1 | 0.5 | 2,744 | 4.6 | 15.3 | 17,990 | 7.1 |
| Ohio | 199 | 5.0 | 1.8 | 2,609 | 4.4 | 24.1 | 10,847 | 4.3 |
| Oklahoma | 83 | 2.1 | 2.6 | 933 | 1.6 | 29.6 | 3,146 | 1.2 |
| Oregon | 69 | 1.7 | 2.4 | 771 | 1.3 | 27.1 | 2,842 | 1.1 |
| Pennsylvania | 117 | 3.0 | 1.0 | 3,573 | 6.0 | 30.1 | 11,882 | 4.7 |
| Puerto Rico | 18 | 0.4 | 0.5 | 996 | 1.7 | 28.3 | 3,522 | 1.4 |
| Rhode Island | 1 | <0.1 | 0.1 | 139 | 0.2 | 13.8 | 1,003 | 0.4 |
| South Carolina | 49 | 1.2 | 1.4 | 1,532 | 2.6 | 43.9 | 3,487 | 1.4 |
| South Dakota | 76 | 1.9 | 10.9 | 272 | 0.5 | 39.1 | 696 | 0.3 |
| Tennessee | 112 | 2.8 | 2.3 | 1,797 | 3.0 | 36.8 | 4,877 | 1.9 |
| Texas | 192 | 4.9 | 1.1 | 3,157 | 5.3 | 18.6 | 16,987 | 6.7 |
| Utah | 12 | 0.3 | 0.7 | 212 | 0.4 | 12.3 | 1,723 | 0.7 |
| Vermont | 12 | 0.3 | 2.1 | 370 | 0.6 | 65.8 | 563 | 0.2 |
| Virginia | 81 | 2.0 | 1.3 | 1,813 | 3.0 | 29.3 | 6,187 | 2.4 |
| Washington | 60 | 1.5 | 1.2 | 1,089 | 1.8 | 22.4 | 4,867 | 1.9 |
| West Virginia | 24 | 0.6 | 1.3 | 1,122 | 1.9 | 62.6 | 1,793 | 0.7 |
| Wisconsin | 196 | 4.9 | 4.0 | 1,484 | 2.5 | 30.3 | 4,892 | 1.9 |
| Wyoming | 16 | 0.4 | 3.5 | 143 | 0.2 | 31.6 | 454 | 0.2 |
| Total | 3,889 | 100.0 | 1.5 | 58,782 | 100.0 | 23.3 | 252,232 | 100.0 |

APPENDIX TABLE 2 Number of Graduates in Agriculture, Food, and Natural Resource Disciplines, by Region and Degree, 1992

| Academic Area | 1862s | 1890s | Non-Land Grant | Total |
|---|---|---|---|---|
| *North-Central* | | | | |
| Associate's Degree Program | | | | |
| General agriculture | 53 | 0 | 0 | 53 |
| Animal sciences | 97 | 0 | 0 | 97 |
| Plant sciences | 165 | 0 | 0 | 165 |
| Soil sciences | 10 | 0 | 0 | 10 |
| Agricultural business and management | 22 | 0 | 0 | 22 |
| Social sciences | 4 | 0 | 0 | 4 |
| Natural resources | 16 | 0 | 0 | 16 |
| Agricultural engineering/mechanization | 13 | 0 | 0 | 13 |
| Food sciences | 0 | 0 | 0 | 0 |
| Related sciences | 0 | 0 | 0 | 0 |
| Other | 17 | 0 | 0 | 17 |
| Subtotal | 397 | 0 | 0 | 397 |
| Bachelor's Degree Program | | | | |
| General agriculture | 137 | 14 | 120 | 271 |
| Animal sciences | 642 | 0 | 123 | 765 |
| Plant sciences | 526 | 0 | 91 | 617 |
| Soil sciences | 25 | 0 | 21 | 46 |
| Agricultural business and management | 663 | 0 | 226 | 889 |
| Social sciences | 236 | 0 | 38 | 274 |
| Natural resources | 727 | 0 | 286 | 1,013 |
| Agricultural engineering/mechanization | 145 | 0 | 24 | 169 |
| Food sciences | 593 | 1 | 55 | 649 |
| Related sciences | 162 | 0 | 9 | 171 |
| Other | 314 | 1 | 31 | 346 |
| Subtotal | 4,170 | 16 | 1,024 | 5,210 |
| Master's Degree Program | | | | |
| General agriculture | 2 | 0 | 6 | 8 |
| Animal sciences | 140 | 0 | 5 | 145 |
| Plant sciences | 231 | 0 | 12 | 243 |
| Soil sciences | 34 | 0 | 3 | 37 |
| Agricultural business and management | 146 | 0 | 11 | 157 |
| Social sciences | 112 | 0 | 8 | 120 |
| Natural resources | 170 | 0 | 29 | 199 |
| Agricultural engineering/mechanization | 42 | 0 | 2 | 44 |
| Food sciences | 155 | 0 | 1 | 156 |
| Related sciences | 77 | 0 | 0 | 77 |
| Other | 18 | 0 | 0 | 18 |
| Subtotal | 11,27 | 0 | 77 | 1,204 |
| Doctorate Degree Program | | | | |
| General agriculture | 0 | 0 | 0 | 0 |
| Animal sciences | 86 | 0 | 0 | 86 |
| Plant sciences | 165 | 0 | 0 | 165 |
| Soil sciences | 29 | 0 | 0 | 29 |
| Agricultural business and management | 40 | 0 | 0 | 40 |
| Social sciences | 47 | 0 | 0 | 47 |
| Natural resources | 52 | 0 | 5 | 57 |

| Academic Area | 1862s | 1890s | Non-Land Grant | Total |
|---|---|---|---|---|
| Doctorate Degree Program | | | | |
| Agricultural engineering/mechanization | 32 | 0 | 0 | 32 |
| Food sciences | 76 | 0 | 0 | 76 |
| Related sciences | 97 | 0 | 0 | 97 |
| Other | 7 | 0 | 0 | 7 |
| Subtotal | 631 | 0 | 5 | 636 |
| *Northeastern* | | | | |
| Associate's Degree Program | | | | |
| General agriculture | 0 | 0 | 0 | 0 |
| Animal sciences | 53 | 0 | 0 | 53 |
| Plant sciences | 161 | 0 | 0 | 161 |
| Soil sciences | 2 | 0 | 0 | 2 |
| Agricultural business and management | 76 | 0 | 0 | 76 |
| Social sciences | 0 | 0 | 0 | 0 |
| Natural resources | 46 | 0 | 0 | 46 |
| Agricultural engineering/mechanization | 0 | 0 | 0 | 0 |
| Food sciences | 17 | 0 | 0 | 17 |
| Related sciences | 0 | 0 | 0 | 0 |
| Other | 26 | 0 | 0 | 26 |
| Subtotal | 381 | 0 | 0 | 381 |
| Bachelor's Degree Program | | | | |
| General agriculture | 116 | 15 | 0 | 131 |
| Animal sciences | 449 | 0 | 0 | 449 |
| Plant sciences | 383 | 1 | 50 | 434 |
| Soil sciences | 12 | 0 | 0 | 12 |
| Agricultural business and management | 574 | 5 | 0 | 579 |
| Social sciences | 174 | 0 | 0 | 174 |
| Natural resources | 588 | 0 | 272 | 860 |
| Agricultural engineering/mechanization | 88 | 0 | 0 | 88 |
| Food sciences | 156 | 0 | 0 | 156 |
| Related sciences | 590 | 0 | 0 | 590 |
| Other | 578 | 4 | 0 | 582 |
| Subtotal | 3,708 | 25 | 322 | 4,055 |
| Master's Degree Program | | | | |
| General agriculture | 0 | 0 | 0 | 0 |
| Animal sciences | 71 | 0 | 0 | 71 |
| Plant sciences | 129 | 0 | 29 | 158 |
| Soil sciences | 8 | 0 | 2 | 10 |
| Agricultural business and management | 54 | 0 | 0 | 54 |
| Social sciences | 73 | 1 | 3 | 77 |
| Natural resources | 53 | 0 | 138 | 191 |
| Agricultural engineering/mechanization | 24 | 0 | 14 | 38 |
| Food sciences | 91 | 0 | 0 | 91 |
| Related sciences | 95 | 0 | 0 | 95 |
| Other | 82 | 1 | 0 | 83 |
| Subtotal | 680 | 2 | 186 | 868 |
| Doctorate Degree Program | | | | |
| General agriculture | 0 | 0 | 0 | 0 |
| Animal sciences | 26 | 0 | 0 | 26 |

| Academic Area | 1862s | 1890s | Non-Land Grant | Total |
|---|---|---|---|---|
| Doctorate Degree Program | | | | |
| Plant sciences | 60 | 0 | 1 | 61 |
| Soil sciences | 1 | 0 | 0 | 1 |
| Agricultural business and management | 20 | 0 | 0 | 20 |
| Social sciences | 15 | 0 | 2 | 17 |
| Natural resources | 14 | 0 | 10 | 24 |
| Agricultural engineering/mechanization | 8 | 0 | 0 | 8 |
| Food sciences | 55 | 0 | 0 | 55 |
| Related sciences | 88 | 0 | 0 | 88 |
| Other | 15 | 0 | 0 | 15 |
| Subtotal | 302 | 0 | 13 | 315 |
| Region total | 5,071 | 27 | 521 | 5,619 |
| *Southern* | | | | |
| Associate's Degree Program | | | | |
| General agriculture | 2 | 0 | 4 | 6 |
| Animal sciences | 16 | 0 | 2 | 18 |
| Plant sciences | 95 | 0 | 5 | 100 |
| Soil sciences | 0 | 0 | 0 | 0 |
| Agricultural business and management | 21 | 0 | 1 | 22 |
| Social sciences | 0 | 0 | 0 | 0 |
| Natural resources | 0 | 0 | 0 | 0 |
| Agricultural engineering/mechanization | 1 | 0 | 0 | 1 |
| Food sciences | 4 | 0 | 4 | 8 |
| Related sciences | 1 | 0 | 0 | 1 |
| Other | 0 | 0 | 0 | 0 |
| Subtotal | 140 | 0 | 16 | 156 |
| Bachelor's Degree Program | | | | |
| General agriculture | 51 | 13 | 122 | 186 |
| Animal sciences | 666 | 30 | 282 | 978 |
| Plant sciences | 442 | 19 | 142 | 603 |
| Soil sciences | 16 | 1 | 7 | 24 |
| Agricultural business and management | 598 | 72 | 368 | 1,038 |
| Social sciences | 174 | 12 | 166 | 352 |
| Natural resources | 871 | 5 | 162 | 1,038 |
| Agricultural engineering/mechanization | 135 | 0 | 55 | 190 |
| Food sciences | 223 | 57 | 14 | 294 |
| Related sciences | 443 | 0 | 4 | 447 |
| Other | 273 | 163 | 139 | 575 |
| Subtotal | 3,892 | 372 | 1,461 | 5,725 |
| Master's Degree Program | | | | |
| General agriculture | 26 | 5 | 46 | 77 |
| Animal sciences | 137 | 9 | 19 | 165 |
| Plant sciences | 151 | 2 | 10 | 163 |
| Soil sciences | 30 | 3 | 2 | 35 |
| Agricultural business and management | 139 | 10 | 13 | 162 |
| Social sciences | 79 | 7 | 25 | 111 |
| Natural resources | 258 | 0 | 97 | 355 |

| Academic Area | 1862s | 1890s | Non-Land Grant | Total |
|---|---|---|---|---|
| Master's Degree Program | | | | |
| Agricultural engineering/mechanization | 36 | 0 | 0 | 36 |
| Food sciences | 118 | 14 | 1 | 133 |
| Related sciences | 51 | 0 | 3 | 54 |
| Other | 31 | 9 | 9 | 49 |
| Subtotal | 1,056 | 59 | 225 | 1,340 |
| Doctorate Degree Program | | | | |
| General agriculture | 0 | 0 | 0 | 0 |
| Animal sciences | 84 | 0 | 4 | 88 |
| Plant sciences | 117 | 1 | 7 | 125 |
| Soil sciences | 17 | 0 | 0 | 17 |
| Agricultural business and management | 40 | 0 | 4 | 44 |
| Social sciences | 16 | 0 | 0 | 16 |
| Natural resources | 71 | 0 | 11 | 82 |
| Agricultural engineering/mechanization | 23 | 0 | 0 | 23 |
| Food sciences | 43 | 0 | 0 | 43 |
| Related sciences | 66 | 0 | 0 | 66 |
| Other | 0 | 0 | 0 | 0 |
| Subtotal | 477 | 1 | 26 | 504 |
| Region total | 5,565 | 432 | 1,728 | 7,725 |
| *Western* | | | | |
| Associate's Degree Program | | | | |
| General agriculture | 0 | 0 | 0 | 0 |
| Animal sciences | 0 | 0 | 15 | 15 |
| Plant sciences | 5 | 0 | 14 | 19 |
| Soil sciences | 0 | 0 | 0 | 0 |
| Agricultural business and management | 0 | 0 | 0 | 0 |
| Social sciences | 0 | 0 | 0 | 0 |
| Natural resources | 0 | 0 | 0 | 0 |
| Agricultural engineering/mechanization | 2 | 0 | 0 | 2 |
| Food sciences | 0 | 0 | 2 | 2 |
| Related sciences | 0 | 0 | 0 | 0 |
| Other | 0 | 0 | 0 | 0 |
| Subtotal | 7 | 0 | 31 | 38 |
| Bachelor's Degree Program | | | | |
| General agriculture | 67 | 0 | 26 | 93 |
| Animal sciences | 355 | 0 | 106 | 461 |
| Plant sciences | 239 | 0 | 127 | 366 |
| Soil sciences | 20 | 0 | 14 | 34 |
| Agricultural business and management | 495 | 0 | 323 | 818 |
| Social sciences | 129 | 0 | 33 | 162 |
| Natural resources | 673 | 0 | 241 | 914 |
| Agricultural engineering/mechanization | 20 | 0 | 26 | 46 |
| Food sciences | 247 | 0 | 117 | 364 |
| Related sciences | 282 | 0 | 0 | 282 |
| Other | 639 | 0 | 257 | 896 |
| Subtotal | 31,66 | 0 | 1,270 | 4,436 |

| Academic Area | 1862s | 1890s | Non-Land Grant | Total |
|---|---|---|---|---|
| Master's Degree Program | | | | |
| General agriculture | 4 | 0 | 39 | 43 |
| Animal sciences | 57 | 0 | 5 | 62 |
| Plant sciences | 127 | 0 | 7 | 134 |
| Soil sciences | 19 | 0 | 1 | 20 |
| Agricultural business and management | 100 | 0 | 11 | 111 |
| Social sciences | 51 | 0 | 0 | 51 |
| Natural resources | 220 | 0 | 65 | 285 |
| Agricultural engineering/mechanization | 16 | 0 | 0 | 16 |
| Food sciences | 66 | 0 | 7 | 73 |
| Related sciences | 62 | 0 | 0 | 62 |
| Other | 61 | 0 | 5 | 66 |
| Subtotal | 783 | 0 | 140 | 923 |
| Doctorate Degree Program | | | | |
| General agriculture | 1 | 0 | 0 | 1 |
| Animal sciences | 30 | 0 | 0 | 30 |
| Plant sciences | 87 | 0 | 0 | 87 |
| Soil sciences | 29 | 0 | 0 | 29 |
| Agricultural business and management | 18 | 0 | 0 | 18 |
| Social sciences | 1 | 0 | 0 | 1 |
| Natural resources | 98 | 0 | 22 | 120 |
| Agricultural engineering/mechanization | 4 | 0 | 0 | 4 |
| Food sciences | 21 | 0 | 0 | 21 |
| Related sciences | 83 | 0 | 0 | 83 |
| Other | 11 | 0 | 0 | 11 |
| Subtotal | 383 | 0 | 22 | 405 |
| Region total | 43,39 | 0 | 1,463 | 5,802 |
| Grand total | 21,300 | 475 | 4,818 | 26,593 |

NOTE: "Natural resources" includes forest sciences.
SOURCE: Data are from USDA Food and Agricultural Education Information System (FAEIS).

APPENDIX TABLE 3 Allocations, by Region, of Extension Staff (full-time equivalent staff years) among Base Programs, 1992

| Location | Agricultural Competitiveness and Profitability | Community Resource and Economic Development | Family Development and Resource Management | 4-H and Youth Development |
|---|---|---|---|---|
| *North-Central* | | | | |
| Illinois | 180.0 (35.4) | 28.0 (5.5) | 61.0 (12.0) | 108.0 (21.3) |
| Indiana | 108.4 (30.3) | 22.3 (6.2) | 25.2 (7.0) | 41.5 (11.6) |
| Iowa | 145.5 (33.1) | 42.0 (9.6) | 60.0 (13.7) | 77.0 (17.5) |
| Kansas | 164.5 (35.0) | 23.5 (5.0) | 42.3 (9.0) | 84.6 (18.0) |
| Michigan | 127.0 (34.6) | 22.0 (6.0) | 42.9 (11.7) | 68.3 (18.6) |
| Minnesota | 114.0 (27.8) | 46.0 (11.2) | 57.0 (13.9) | 86.0 (21.0) |
| Missouri | 127.0 (31.5) | 52.5 (13.0) | 55.3 (13.7) | 63.0 (15.6) |
| North Dakota | 76.5 (40.3) | 9.9 (5.2) | 21.3 (11.2) | 36.4 (19.2) |
| Nebraska | 81.7 (37.0) | 8.8 (4.0) | 17.7 (8.0) | 33.1 (15.0) |
| Ohio | 97.0 (25.7) | 17.0 (4.5) | 60.0 (15.9) | 110.0 (29.2) |
| South Dakota | 80.0 (47.4) | 0.0 (0.0) | 14.7 (8.7) | 46.0 (27.3) |
| Wisconsin | 141.4 (29.2) | 70.0 (14.5) | 72.8 (15.0) | 75.8 (15.7) |
| Subtotal | 1,443.0 (32.8) | 342.0 (7.8) | 530.2 (12.1) | 829.7 (18.9) |
| *Northeastern* | | | | |
| Connecticut | 23.0 (27.4) | 1.5 (1.8) | 8.0 (9.5) | 12.0 (14.3) |
| District of Columbia | 0.0 (0.0) | 0.0 (0.0) | 0.9 (4.7) | 7.4 (38.3) |
| Delaware | 20.2 (41.6) | 0.0 (0.0) | 3.5 (7.2) | 5.9 (12.1) |
| Massachusetts | 31.8 (26.9) | 5.0 (4.2) | 10.0 (8.5) | 18.5 (15.7) |
| Maryland | 53.0 (25.6) | 3.9 (1.9) | 19.0 (9.2) | 14.5 (7.0) |
| Maine | 32.6 (28.7) | 4.7 (4.1) | 7.1 (6.2) | 23.0 (20.2) |
| New Hampshire | 26.2 (26.1) | 2.0 (2.0) | 12.0 (12.0) | 18.0 (18.0) |
| New Jersey | 46.3 (30.0) | 20.0 (13.0) | 9.3 (6.0) | 16.9 (11.0) |
| New York | 205.7 (30.0) | 34.3 (5.0) | 61.7 (9.0) | 150.9 (22.0) |
| Pennsylvania | 170.7 (38.0) | 16.8 (3.7) | 74.1 (16.5) | 76.5 (17.0) |
| Rhode Island | 13.7 (39.1) | 2.7 (7.7) | 1.7 (4.9) | 4.8 (13.7) |
| Vermont | 19.7 (26.1) | 6.1 (8.1) | 5.5 (7.3) | 6.8 (9.0) |
| West Virginia | 36.1 (19.0) | 26.6 (14.0) | 15.2 (8.0) | 57.0 (30.0) |
| Subtotal | 679.0 (29.8) | 123.6 (5.4) | 228.0 (10.0) | 412.2 (18.1) |
| *Southern* | | | | |
| Alabama | 115.1 (27.6) | 34.1 (8.2) | 66.5 (16.0) | 90.5 (21.7) |
| Arkansas | 166.2 (40.0) | 20.8 (5.0) | 62.3 (15.0) | 103.9 (25.0) |
| Florida | 149.0 (33.0) | 6.0 (1.3) | 44.0 (9.8) | 90.0 (20.0) |
| Georgia | 253.9 (39.9) | 24.4 (3.8) | 32.9 (5.2) | 126.5 (19.9) |
| Kentucky | 149.0 (28.9) | 31.0 (6.0) | 84.0 (16.3) | 64.0 (12.4) |
| Louisiana | 125.0 (31.5) | 7.1 (1.8) | 36.3 (9.1) | 96.7 (24.4) |
| Mississippi | 190.0 (44.7) | 15.0 (3.5) | 50.0 (11.8) | 76.0 (17.9) |
| North Carolina | 269.8 (35.8) | 44.8 (5.9) | 135.0 (17.9) | 135.0 (17.9) |
| Oklahoma | 117.0 (35.0) | 20.0 (6.0) | 36.7 (11.0) | 66.8 (20.0) |
| Puerto Rico | 95.0 (39.1) | 21.3 (8.8) | 39.1 (16.1) | 49.5 (20.4) |
| South Carolina | 141.9 (37.5) | 18.2 (4.8) | 54.7 (14.5) | 65.4 (17.3) |
| Tennessee | 180.0 (37.9) | 15.0 (3.2) | 40.0 (8.4) | 93.0 (19.6) |
| Texas | 367.0 (35.9) | 46.5 (4.5) | 134.8 (13.2) | 116.6 (11.4) |
| Virginia | 187.2 (36.5) | 23.6 (4.6) | 64.0 (12.5) | 149.0 (29.1) |
| Virgin Islands | 8.2 (31.5) | 3.0 (11.5) | 2.2 (8.5) | 2.2 (8.5) |
| Subtotal | 2,514.3 (35.9) | 330.8 (4.7) | 882.5 (12.6) | 1,325.10 (18.9) |
| *Western* | | | | |
| Alaska | 7.6 (28.1) | 4.0 (14.8) | 3.0 (11.1) | 5.5 (20.4) |
| American Samoa | 4.6 (34.3) | 0.8 (6.0) | 1.6 (11.9) | 2.5 (18.7) |
| Arizona | 40.0 (28.0) | 15.0 (10.5) | 11.0 (7.7) | 29.0 (20.3) |
| California | 227.5 (54.3) | 14.4 (3.4) | 14.4 (3.4) | 46.5 (11.1) |

| Location | Leadership and Volunteer Development | Natural Resources and Environment Management | Nutrition, Diet, and Health | Other | Total |
| --- | --- | --- | --- | --- | --- |
| *North-Central* | | | | | |
| Illinois | 47.0 (9.3) | 35.0 (6.9) | 49.0 (9.6) | 0.0 (0.0) | 508.0 |
| Indiana | 59.0 (16.5) | 65.4 (18.3) | 28.0 (7.8) | 8.2 (2.3) | 358.0 |
| Iowa | 22.0 (5.0) | 55.0 (12.5) | 33.0 (7.5) | 5.0 (1.1) | 439.5 |
| Kansas | 37.6 (8.0) | 23.5 (5.0) | 51.7 (11.0) | 42.3 (9.0) | 470.0 |
| Michigan | 30.5 (8.3) | 40.7 (11.1) | 35.6 (9.7) | 0.0 (0.0) | 367.0 |
| Minnesota | 32.0 (7.8) | 44.0 (10.7) | 31.0 (7.6) | 0.0 (0.0) | 410.0 |
| Missouri | 44.0 (10.9) | 48.0 (11.9) | 13.0 (3.2) | 0.0 (0.0) | 402.8 |
| North Dakota | 9.9 (5.2) | 21.7 (11.4) | 14.3 (7.5) | 0.0 (0.0) | 190.0 |
| Nebraska | 22.1 (10.0) | 28.7 (13.0) | 28.7 (13.0) | 0.0 (0.0) | 220.8 |
| Ohio | 25.0 (6.6) | 20.0 (5.3) | 30.0 (8.0) | 18.0 (4.8) | 377.0 |
| South Dakota | 5.8 (3.4) | 7.0 (4.1) | 15.3 (9.1) | 0.0 (0.0) | 168.8 |
| Wisconsin | 15.1 (3.1) | 89.9 (18.6) | 18.8 (3.9) | 0.0 (0.0) | 483.8 |
| Subtotal | 350.0 (8.0) | 478.9 (10.9) | 348.4 (7.9) | 73.5 (1.7) | 4,395.7 |
| *Northeastern* | | | | | |
| Connecticut | 8.0 (9.5) | 19.5 (23.2) | 6.0 (7.1) | 6.0 (7.1) | 84.0 |
| District of Columbia | 2.4 (12.4) | 4.1 (21.2) | 4.5 (23.3) | 0.0 (0.0) | 19.3 |
| Delaware | 2.8 (5.8) | 7.6 (15.6) | 8.6 (17.7) | 0.0 (0.0) | 48.6 |
| Massachusetts | 14.0 (11.9) | 20.2 (17.1) | 12.5 (10.6) | 6.0 (5.1) | 118.0 |
| Maryland | 17.2 (8.3) | 24.6 (11.9) | 15.8 (7.6) | 59.0 (28.5) | 207.0 |
| Maine | 3.6 (3.2) | 30.9 (27.2) | 7.5 (6.6) | 4.3 (3.8) | 113.7 |
| New Hampshire | 8.5 (8.5) | 21.5 (21.5) | 12.0 (12.0) | 0.0 (0.0) | 100.2 |
| New Jersey | 6.2 (4.0) | 20.0 (13.0) | 35.5 (23.0) | 0.0 (0.0) | 154.2 |
| New York | 61.7 (9.0) | 68.6 (10.0) | 102.8 (15.0) | 0.0 (0.0) | 685.7 |
| Pennsylvania | 32.9 (7.3) | 30.6 (6.8) | 47.1 (10.5) | 0.0 (0.0) | 448.7 |
| Rhode Island | 2.6 (7.4) | 3.4 (9.7) | 5.1 (14.6) | 1.0 (2.9) | 35.0 |
| Vermont | 5.7 (7.6) | 9.6 (12.7) | 11.5 (15.3) | 10.5 (13.9) | 75.4 |
| West Virginia | 22.8 (12.0) | 15.2 (8.0) | 17.1 (9.0) | 0.0 (0.0) | 190.0 |
| Subtotal | 188.4 (8.3) | 275.8 (12.1) | 286.0 (12.5) | 86.8 (3.8) | 2,279.8 |
| *Southern* | | | | | |
| Alabama | 19.9 (4.8) | 44.9 (10.8) | 45.3 (10.9) | 0.0 (0.0) | 416.3 |
| Arkansas | 20.8 (5.0) | 20.8 (5.0) | 20.8 (5.0) | 0.0 (0.0) | 415.6 |
| Florida | 16.0 (3.5) | 101.0 (22.4) | 45.0 (10.0) | 0.0 (0.0) | 451.0 |
| Georgia | 64.0 (10.1) | 82.4 (13.0) | 51.4 (8.1) | 0.7 (0.1) | 636.2 |
| Kentucky | 83.0 (16.1) | 53.0 (10.3) | 52.0 (10.1) | 0.0 (0.0) | 516.0 |
| Louisiana | 64.7 (16.3) | 21.8 (5.5) | 45.4 (11.4) | 0.0 (0.0) | 397.0 |
| Mississippi | 28.0 (6.6) | 12.0 (2.8) | 54.0 (12.7) | 0.0 (0.0) | 425.0 |
| North Carolina | 36.0 (4.8) | 68.4 (9.1) | 65.0 (8.6) | 0.0 (0.0) | 754.0 |
| Oklahoma | 33.4 (10.0) | 33.4 (10.0) | 26.7 (8.0) | 0.0 (0.0) | 334.0 |
| Puerto Rico | 17.4 (7.2) | 1.8 (0.7) | 18.7 (7.7) | 0.0 (0.0) | 242.8 |
| South Carolina | 20.4 (5.4) | 50.0 (13.2) | 27.9 (7.4) | 0.0 (0.0) | 378.5 |
| Tennessee | 40.0 (8.4) | 27.0 (5.7) | 52.0 (10.9) | 28.4 (6.0) | 475.4 |
| Texas | 72.7 (7.1) | 128.2 (12.5) | 157.2 (15.4) | 0.0 (0.0) | 1,023.0 |
| Virginia | 26.8 (5.2) | 27.6 (5.4) | 34.3 (6.7) | 0.0 (0.0) | 512.5 |
| Virgin Islands | 2.2 (8.5) | 5.7 (21.9) | 2.5 (9.6) | 0.0 (0.0) | 26.0 |
| Subtotal | 545.3 (7.8) | 678.0 (9.7) | 698.2 (10.0) | 29.1 (0.4) | 7,003.3 |
| *Western* | | | | | |
| Alaska | 1.5 (5.6) | 1.5 (5.6) | 3.9 (14.4) | 0.0 (0.0) | 27.0 |
| American Samoa | 1.1 (8.2) | 1.5 (11.2) | 1.3 (9.7) | 0.0 (0.0) | 13.4 |
| Arizona | 15.0 (10.5) | 26.0 (18.2) | 7.0 (4.9) | 0.0 (0.0) | 143.0 |
| California | 15.5 (3.7) | 52.6 (12.6) | 26.8 (6.4) | 20.9 (5.0) | 418.6 |

| Location | Community Agricultural Competitiveness and Profitability | Family Resource and Economic Development | Development and Resource Management | 4-H and Youth Development |
|---|---|---|---|---|
| Colorado | 54.3 (25.5) | 7.7 (3.6) | 15.3 (7.2) | 44.9 (21.1) |
| Guam | 0.0 (0.0) | 3.0 (15.8) | 3.5 (18.4) | 2.0 (10.5) |
| Hawaii | 62.8 (58.7) | 0.0 (0.0) | 7.2 (6.7) | 11.0 (10.3) |
| Idaho | 67.0 (43.5) | 4.0 (2.6) | 9.1 (5.9) | 27.6 (17.9) |
| Micronesia | 1.3 (10.5) | 0.4 (3.2) | 1.8 (14.5) | 3.2 (25.8) |
| Montana | 52.0 (19.4) | 35.0 (13.1) | 38.0 (14.2) | 40.0 (14.9) |
| Nevada | 14.0 (20.3) | 1.0 (1.4) | 2.0 (2.9) | 14.0 (20.3) |
| New Mexico | 37.0 (29.0) | 7.4 (5.8) | 17.9 (14.0) | 24.9 (19.5) |
| Northern Marianas | 1.6 (18.0) | 0.3 (3.4) | 4.3 (48.3) | 0.4 (4.5) |
| Oregon | 95.1 (36.4) | 10.6 (4.1) | 20.6 (7.9) | 17.0 (6.5) |
| Utah | 34.0 (25.6) | 10.0 (7.5) | 16.0 (12.0) | 27.0 (20.3) |
| Washington | 83.0 (37.9) | 15.3 (7.0) | 35.5 (16.2) | 26.3 (12.0) |
| Wyoming | 34.6 (34.8) | 5.9 (5.9) | 11.7 (11.8) | 18.4 (18.5) |
| Subtotal | 816.4 (35.6) | 134.8 (5.9) | 212.9 (9.3) | 340.2 (14.8) |
| Total | 5,452.7 (34.1) | 931.2 (5.8) | 1,853.6 (11.6) | 2,907.2 (18.2) |

NOTE: Number in parentheses is percent of full-time equivalent staff years.
SOURCE: Data were provided by the USDA Extension Service.

| Location | Leadership and Volunteer Development | Natural Resources and Environment Management | Nutrition, Diet, and Health | Other | Total |
|---|---|---|---|---|---|
| Colorado | 38.7 (18.2) | 19.8 (9.3) | 32.1 (15.1) | 0.0 (0.0) | 212.8 |
| Guam | 1.0 (5.3) | 7.0 (36.8) | 2.5 (13.2) | 0.0 (0.0) | 19.0 |
| Hawaii | 9.1 (8.5) | 10.2 (9.5) | 6.6 (6.2) | 0.0 (0.0) | 106.9 |
| Idaho | 12.5 (8.1) | 19.6 (12.7) | 14.2 (9.2) | 0.0 (0.0) | 154.0 |
| Micronesia | 1.8 (14.5) | 1.0 (8.1) | 2.9 (23.4) | 0.0 (0.0) | 12.4 |
| Montana | 37.0 (13.8) | 36.0 (13.4) | 30.0 (11.2) | 0.0 (0.0) | 268.0 |
| Nevada | 7.0 (10.1) | 14.0 (20.3) | 4.0 (5.8) | 13.0 (18.8) | 69.0 |
| New Mexico | 22.5 (17.6) | 11.7 (9.2) | 6.1 (4.8) | 0.0 (0.0) | 127.5 |
| Northern Marianas | 0.0 (0.0) | 0.5 (5.6) | 1.8 (20.2) | 0.0 (0.0) | 8.9 |
| Oregon | 27.8 (10.6) | 25.2 (9.6) | 18.4 (7.0) | 46.9 (17.9) | 261.6 |
| Utah | 10.0 (7.5) | 14.0 (10.5) | 13.0 (9.8) | 9.0 (6.8) | 133.0 |
| Washington | 9.0 (4.1) | 36.8 (16.8) | 13.1 (6.0) | 0.0 (0.0) | 219.0 |
| Wyoming | 8.2 (8.3) | 10.9 (11.0) | 9.6 (9.7) | 0.0 (0.0) | 99.3 |
| Subtotal | 217.7 (9.5) | 288.3 (12.6) | 193.3 (8.4) | 89.8 (3.9) | 2,293.4 |
| Total | 1,301.4 (8.1) | 1,721.0 (10.8) | 1,525.90 (9.6) | 279.2 (1.7) | 15,972.2 |

APPENDIX TABLE 4 Allocation of Research Expenditures (thousands of dollars) Among CRIS Research Program Groups, by Institution, 1992

| Institution | Natural Resources | Forest Resources | Crops | Animals |
|---|---|---|---|---|
| *1862 State Agricultural Experiment Stations* | | | | |
| U. of California | 19,604 (13.2) | 8,735 (5.9) | 61,673 (41.6) | 23,118 (15.6) |
| Texas A&M U. | 10,444 (11.1) | 4,541 (4.8) | 41,005 (43.7) | 24,421 (26.0) |
| U. of Florida | 10,068 (11.3) | 6,288 (7.1) | 43,531 (48.9) | 17,796 (20.0) |
| Cornell U. | 9,674 (12.8) | 1,595 (2.1) | 26,041 (34.4) | 10,885 (14.4) |
| U. of Wisconsin | 5,276 (8.0) | 2,758 (4.2) | 13,173 (19.9) | 18,245 (27.6) |
| North Carolina State U. | 6,107 (9.7) | 921 (1.5) | 28,108 (44.6) | 15,479 (24.6) |
| Michigan State U. | 10,623 (17.6) | 4,520 (7.5) | 17,791 (29.5) | 10,800 (17.9) |
| Iowa State U. | 8,006 (13.4) | 1,721 (2.9) | 16,664 (28.0) | 16,334 (27.4) |
| U. of Minnesota | 7,135 (12.3) | 6,385 (11.0) | 18,025 (31.0) | 13,555 (23.3) |
| U. of Nebraska | 7,587 (14.6) | 1,011 (1.9) | 15,163 (29.2) | 21,242 (41.0) |
| Purdue U. | 4,661 (9.1) | 2,461 (4.8) | 20,252 (39.5) | 13,094 (25.5) |
| U. of Georgia | 4,010 (8.0) | 5,410 (10.8) | 23,401 (46.7) | 10,121 (20.2) |
| Virginia Polytechnic Institute and State U. | 6,824 (15.7) | 6,765 (15.5) | 9,752 (22.4) | 12,340 (28.3) |
| Oregon State U. | 7,557 (18.0) | 1,589 (3.8) | 17,601 (41.8) | 7,794 (18.5) |
| Kansas State U. | 3,659 (8.8) | 632 (1.5) | 20,201 (48.6) | 12,491 (30.1) |
| U. of Arizona | 6,060 (15.3) | 3,198 (8.0) | 11,919 (30.0) | 9,862 (24.8) |
| Louisiana State U. | 2,232 (5.9) | 2,250 (5.9) | 17,510 (46.1) | 11,140 (29.3) |
| U. of Illinois | 2,746 (7.3) | 1,651 (4.4) | 13,587 (36.0) | 12,050 (32.0) |
| Ohio State U. | 5,900 (15.7) | 2,205 (5.9) | 13,184 (35.0) | 10,297 (27.4) |
| Pennsylvania State U. | 3,894 (10.6) | 3,706 (10.1) | 10,777 (29.3) | 10,680 (29.0) |
| Washington State U. | 4,035 (11.0) | 1,578 (4.3) | 20,635 (56.2) | 5,506 (15.0) |
| Mississippi State U. | 3,065 (8.6) | 7,105 (20.0) | 11,397 (32.1) | 10,980 (31.0) |
| U. of Missouri | 3,971 (11.7) | 1,736 (5.1) | 10,715 (31.5) | 9,757 (28.7) |
| Auburn U. | 5,341 (16.5) | 3,829 (11.8) | 8,792 (27.1) | 9,999 (30.8) |
| U. of Arkansas | 3,281 (10.2) | 2,211 (6.8) | 15,254 (47.2) | 6,783 (21.0) |
| Colorado State U. | 10,851 (36.3) | 1,701 (5.7) | 9,562 (32.0) | 4,499 (15.1) |
| Clemson U. | 1,197 (4.2) | 49 (0.2) | 14,539 (50.8) | 7,957 (27.8) |
| Oklahoma State U. | 2,741 (9.7) | 1,397 (5.0) | 10,626 (37.7) | 9,810 (34.8) |
| North Dakota State U. | 2,779 (10.6) | 875 (3.3) | 13,670 (51.9) | 5,316 (20.2) |
| U. of Kentucky | 2,184 (8.5) | 1,672 (6.5) | 9,898 (38.5) | 7,839 (30.5) |
| U. of Tennessee | 2,681 (10.7) | 2,057 (8.2) | 9,607 (38.2) | 8,201 (32.6) |
| Rutgers—The State U., Cook College | 4,090 (16.3) | 724 (2.9) | 5,985 (23.9) | 6,340 (25.3) |
| U. of Hawaii | 2,332 (10.0) | 997 (4.3) | 13,763 (59.3) | 3,097 (13.3) |
| U. of Marvland | 3,886 (18.6) | 1,472 (7.1) | 6,761 (32.4) | 6,224 (29.8) |
| U. of Idaho | 1,098 (10.5) | 597 (3.3) | 9,845 (54.1) | 2,424 (13.3) |
| Utah State U. | 2,721 (16.0) | 1,093 (6.4) | 4,536 (26.7) | 6,403 (37.7) |
| Montana State U. | 1,267 (7.9) | 894 (5.6) | 7,575 (47.4) | 4,696 (29.4) |
| South Dakota State U. | 1,980 (13.7) | 326 (2.3) | 4,513 (31.2) | 6,900 (47.6) |
| New Mexico State U. | 1,917 (14.8) | 923 (7.1) | 6,487 (50.0) | 1,932 (14.9) |
| U. of Maine | 1,759 (14.3) | 3,630 (29.4) | 3,905 (31.7) | 1,753 (14.2) |
| U. of Massachusetts | 1,172 (10.1) | 1,130 (9.7) | 2,806 (24.2) | 2,503 (21.6) |
| U. of Puerto Rico | 489 (4.3) | 150 (1.3) | 9,261 (81.6) | 890 (7.8) |
| U. of Connecticut | 1,566 (14.0) | 968 (8.7) | 3,232 (28.9) | 2,956 (26.4) |
| U. of Delaware | 1,646 (17.4) | 277 (2.9) | 2,653 (28.1) | 3,928 (41.6) |
| West Virginia U. | 1,779 (21.6) | 2,315 (28.1) | 1,380 (16.8) | 2,355 (28.6) |
| U. of Nevada | 1,896 (25.2) | 497 (6.6) | 1,120 (14.9) | 1,886 (25.1) |
| U. of Vermont | 808 (11.2) | 482 (6.7) | 1,610 (22.4) | 1,433 (19.9) |
| U. of Wyoming | 402 (6.4) | 488 (7.8) | 1,266 (20.2) | 2,952 (47.1) |
| U. of Alaska | 2,105 (37.0) | 1,078 (19.0) | 1,376 (24.2) | 929 (16.3) |
| U. of New Hampshire | 601 (12.8) | 740 (15.8) | 1,097 (23.4) | 1,787 (38.2) |
| U. of Rhode Island | 1,056 (29.5) | 280 (7.8) | 842 (23.6) | 397 (11.1) |

| People, Communities, and Institutions, including Rural Development | | Competition, Trade Adjustment, Price, and Income Policy | | General Resource or Technology | | Food Science and Human Nutrition | | Total Research Expenditures |
|---|---|---|---|---|---|---|---|---|
| 7,990 | (5.4) | 3,826 | (2.6) | 13,704 | (9.2) | 9,564 | (6.5) | 148,214 |
| 2,238 | (2.4) | 4,153 | (4.4) | 4,305 | (4.6) | 2,803 | (3.0) | 93,910 |
| 2,835 | (3.2) | 2,189 | (2.5) | 2,413 | (2.7) | 3,960 | (4.4) | 89,081 |
| 6,653 | (8.8) | 3,242 | (4.3) | 8,372 | (11.0) | 9,342 | (12.3) | 75,805 |
| 1,893 | (2.9) | 5,129 | (7.7) | 14,220 | (21.5) | 5,516 | (8.3) | 66,210 |
| 1,045 | (1.7) | 4,858 | (7.7) | 4,497 | (7.1) | 2,028 | (3.2) | 63,044 |
| 1,841 | (3.1) | 6,426 | (10.7) | 2,492 | (4.1) | 5,811 | (9.6) | 60,304 |
| 6,322 | (10.6) | 6,534 | (11.0) | 1,786 | (3.0) | 2,222 | (3.7) | 59,588 |
| 4,069 | (7.0) | 3,262 | (5.6) | 1,839 | (3.2) | 3,952 | (6.8) | 58,221 |
| 2,567 | (4.9) | 1,347 | (2.6) | 1,199 | (2.3) | 1,743 | (3.4) | 51,858 |
| 950 | (1.9) | 2,545 | (5.0) | 4,795 | (9.3) | 2,524 | (4.9) | 51,281 |
| 1,549 | (3.1) | 1,666 | (3.3) | 828 | (1.7) | 3,134 | (6.3) | 50,119 |
| 2,002 | (4.6) | 952 | (2.2) | 3,085 | (7.1) | 1,810 | (4.2) | 43,530 |
| 943 | (2.2) | 2,024 | (4.8) | 1,309 | (3.1) | 3,268 | (7.8) | 42,085 |
| 798 | (1.9) | 801 | (1.9) | 1,166 | (2.8) | 1,792 | (4.3) | 41,540 |
| 3,101 | (7.8) | 2,026 | (5.1) | 1,707 | (4.3) | 1,858 | (4.7) | 39,731 |
| 2,500 | (6.6) | 777 | (2.0) | 373 | (1.0) | 1,232 | (3.2) | 38,014 |
| 1,537 | (4.1) | 2,326 | (6.2) | 1,606 | (4.3) | 2,208 | (5.9) | 37,711 |
| 1,729 | (4.6) | 2,106 | (5.6) | 417 | (1.1) | 1,798 | (4.8) | 37,636 |
| 2,405 | (6.5) | 2,573 | (7.0) | 546 | (1.5) | 2,185 | (5.9) | 36,767 |
| 1,029 | (2.8) | 1,681 | (4.6) | 1,286 | (3.5) | 986 | (2.7) | 36,735 |
| 962 | (2.7) | 716 | (2.0) | 197 | (0.6) | 1,040 | (2.9) | 35,462 |
| 1,889 | (5.6) | 2,382 | (7.0) | 2,065 | (6.1) | 1,478 | (4.3) | 33,993 |
| 1,519 | (4.7) | 1,080 | (3.3) | 1,061 | (3.3) | 829 | (2.6) | 32,451 |
| 650 | (2.0) | 1,822 | (5.6) | 807 | (2.5) | 1,506 | (4.7) | 32,315 |
| 381 | (1.3) | 482 | (1.6) | 638 | (2.1) | 1,777 | (5.9) | 29,892 |
| 619 | (2.2) | 1,645 | (5.7) | 1,259 | (4.4) | 1,378 | (4.8) | 28,644 |
| 336 | (1.2) | 2,086 | (7.4) | 865 | (3.1) | 332 | (1.2) | 28,192 |
| 517 | (2.0) | 2,164 | (8.2) | 630 | (2.4) | 382 | (1.4) | 26,334 |
| 1,011 | (3.9) | 1,774 | (6.9) | 1,029 | (4.0) | 276 | (1.1) | 25,682 |
| 341 | (1.4) | 936 | (3.7) | 165 | (0.7) | 149 | (4.6) | 25,137 |
| 938 | (3.7) | 781 | (3.1) | 2,518 | (10.1) | 3,671 | (14.7) | 25,046 |
| 527 | (2.3) | 929 | (4.0) | 837 | (3.6) | 744 | (3.2) | 23,224 |
| 227 | (1.1) | 920 | (4.4) | 838 | (4.0) | 547 | (2.6) | 20,875 |
| 497 | (2.7) | 1,258 | (6.9) | 699 | (3.8) | 963 | (5.3) | 18,191 |
| 695 | (4.1) | 280 | (1.6) | 254 | (1.5) | 1,021 | (6.0) | 17,002 |
| 316 | (2.0) | 374 | (2.3) | 592 | (3.7) | 257 | (1.6) | 15,971 |
| 88 | (0.6) | 287 | (2.0) | 154 | (1.1) | 232 | (1.6) | 14,480 |
| 614 | (4.7) | 453 | (3.5) | 444 | (3.4) | 211 | (1.6) | 12,981 |
| 361 | (2.9) | 603 | (4.9) | 3 | (0.0) | 318 | (2.6) | 12,331 |
| 783 | (6.7) | 152 | (1.3) | 564 | (4.9) | 2,492 | (21.5) | 11,602 |
| 148 | (1.3) | 262 | (2.3) | 0 | (0.0) | 154 | (1.4) | 11,353 |
| 561 | (5.0) | 542 | (4.8) | 0 | (0.0) | 1,355 | (12.1) | 11,181 |
| 249 | (2.6) | 404 | (4.3) | 10 | (0.1) | 277 | (2.9) | 9,445 |
| 233 | (2.8) | 89 | (1.1) | 0 | (0.0) | 75 | (0.9) | 8,226 |
| 482 | (6.4) | 384 | (5.1) | 1,124 | (14.9) | 137 | (1.8) | 7,526 |
| 975 | (13.6) | 276 | (3.8) | 128 | (1.8) | 1,475 | (20.5) | 7,187 |
| 399 | (6.4) | 407 | (6.5) | 16 | (0.3) | 331 | (5.3) | 6,260 |
| 39 | (0.7) | 139 | (2.5) | 18 | (0.3) | 0 | (0.0) | 5,685 |
| 182 | (3.9) | 39 | (0.8) | 162 | (3.5) | 73 | (1.6) | 4,681 |
| 373 | (10.4) | 116 | (3.3) | 29 | (0.8) | 483 | (13.5) | 3,577 |

| Institution | Natural Resources | Forest Resources | Crops | Animals |
|---|---|---|---|---|
| U. of Guam | 185 (6.8) | 74 (2.7) | 1,997 (73.3) | 299 (11.0) |
| U. of the Virgin Islands | 82 (5.8) | 94 (6.6) | 770 (54.5) | 467 (33.0) |
| U. of the District of Columbia | 264 (32.8) | 58 (7.3) | 129 (16.0) | 6 (0.7) |
| American Samoa Community College | 0 (0.0) | 0 (0.0) | 641 (100.0) | 0 (0.0) |
| Northern Marianas College | 0 (0.0) | 0 (0.0) | 575 (100.0) | 0 (0.0) |
| Subtotal | 220,106 (11.1) | 111,836 (5.6) | 668,177 (33.5) | 430,947 (21.6) |
| *1890 Colleges and Universities* | | | | |
| Prairie View A&M U. | 462 (15.0) | 0 (0.0) | 111 (3.6) | 1,860 (60.3) |
| North Carolina A&T State U. | 278 (10.5) | 0 (0.0) | 652 (24.6) | 763 (28.8) |
| Kentucky State College | 327 (14.2) | 170 (7.4) | 772 (33.6) | 513 (22.3) |
| Lincoln U. | 20 (0.9) | 0 (0.0) | 584 (27.3) | 688 (32.2) |
| Alabama A&M U. | 180 (8.6) | 30 (1.4) | 845 (40.3) | 200 (9.6) |
| Fort Valley State College | 84 (4.0) | 0 (0.0) | 589 (28.2) | 1,037 (49.6) |
| Tuskegee U. | 105 (5.1) | 0 (0.0) | 885 (42.7) | 365 (17.6) |
| Tennessee State U. | 58 (2.9) | 204 (10.1) | 694 (34.6) | 876 (43.6) |
| South Carolina State College | 0 (0.0) | 0 (0.0) | 165 (10.1) | 0 (0.0) |
| Virginia State U. | 182 (11.3) | 0 (0.0) | 1,005 (62.5) | 297 (18.5) |
| Alcorn A&M U. | 0 (0.0) | 0 (0.0) | 729 (46.7) | 233 (14.9) |
| U. of Arkansas | 13 (0.8) | 0 (0.0) | 383 (24.9) | 729 (47.5) |
| Southern U. | 0 (0.0) | 0 (0.0) | 358 (30.3) | 427 (36.1) |
| U. of Maryland, Eastern Shore | 0 (0.0) | 0 (0.0) | 571 (49.1) | 295 (25.4) |
| Florida A&M U. | 323 (29.4) | 154 (14.0) | 203 (18.5) | 338 (30.8) |
| Langston U. | 18 (1.7) | 0 (0.0) | 193 (17.8) | 875 (80.6) |
| Delaware State U. | 191 (29.1) | 0 (0.0) | 396 (60.2) | 70 (10.7) |
| Subtotal | 2,241 (0.1) | 557 (0.0) | 9,137 (0.5) | 9,565 (0.5) |
| *Forestry Schools* | | | | |
| State U. of New York | 5,070 (27.7) | 12,021 (65.7) | 104 (0.6) | 169 (0.9) |
| Oregon State U. | 1,168 (9.8) | 10,402 (86.9) | 233 (1.9) | 0 (0.0) |
| North Carolina State U. | 730 (10.5) | 6,197 (89.3) | 0 (0.0) | 0 (0.0) |
| U. of Washington | 234 (3.6) | 6,143 (94.3) | 19 (0.3) | 25 (0.4) |
| Michigan Technological U. | 120 (2.6) | 4,540 (97.4) | 0 (0.0) | 0 (0.0) |
| U. of Montana | 132 (4.5) | 2,657 (90.4) | 81 (2.8) | 0 (0.0) |
| U. of Idaho | 490 (16.8) | 2,204 (75.6) | 133 (4.6) | 0 (0.0) |
| Stephen F. Austin State U. | 88 (3.6) | 2,319 (96.4) | 0 (0.0) | 0 (0.0) |
| Clemson U. | 132 (7.3) | 1,644 (90.2) | 0 (0.0) | 7 (0.4) |
| Northern Arizona U. | 0 (0.0) | 1,327 (94.5) | 0 (0.0) | 0 (0.0) |
| U. of Michigan | 0 (0.0) | 1,175 (96.0) | 10 (0.9) | 0 (0.0) |
| U. of Vermont | 62 (6.4) | 911 (93.6) | 0 (0.0) | 0 (0.0) |
| Colorado State U. | 365 (40.5) | 536 (59.5) | 0 (0.0) | 0 (0.0) |
| Southern Illinois U. | 41 (6.3) | 616 (93.7) | 0 (0.0) | 0 (0.0) |
| Louisiana Tech U. | 0 (0.0) | 279 (100.0) | 0 (0.0) | 0 (0.0) |
| Humboldt State U. | 0 (0.0) | 245 (100.0) | 0 (0.0) | 0 (0.0) |
| California Polytechnic State U. | 0 (0.0) | 201 (100.0) | 0 (0.0) | 0 (0.0) |
| Subtotal | 8,634 (0.4) | 53,418 (2.7) | 580 (0.0) | 200 (0.0) |
| *Schools of Veterinary Medicine* | | | | |
| U. of California | 3,005 (11.0) | 0 (0.0) | 0 (0.0) | 24,217 (88.3) |
| Colorado State U. | 0 (0.0) | 0 (0.0) | 0 (0.0) | 22,447 (100.0) |
| Cornell U. | 939 (5.4) | 0 (0.0) | 0 (0.0) | 14,021 (80.0) |
| U. of Illinois | 579 (6.7) | 0 (0.0) | 0 (0.0) | 7,833 (91.1) |

| People, Communities, and Institutions, including Rural Development | | Competition, Trade Adjustment, Price, and Income Policy | | General Resource or Technology | | Food Science and Human Nutrition | | Total Research Expenditures |
|---|---|---|---|---|---|---|---|---|
| 0 | (0.0) | 158 | (5.8) | 10 | (0.4) | 0 | (0.0) | 2,724 |
| 0 | (0.0) | 0 | (0.0) | 0 | (0.0) | 0 | (0.0) | 1,413 |
| 121 | (15.0) | 0 | (0.0) | 35 | (4.3) | 192 | (23.9) | 805 |
| 0 | (0.0) | 0 | (0.0) | 0 | (0.0) | 0 | (0.0) | 641 |
| 0 | (0.0) | 0 | (0.0) | 0 | (0.0) | 0 | (0.0) | 575 |
| 73,032 | (3.7) | 84,384 | (4.2) | 89,096 | (4.5) | 94,889 | (4.8) | 1,772,467 |
| | | | | | | | | |
| 167 | (5.4) | 318 | (10.3) | 0 | (0.0) | 167 | (5.4) | 3,085 |
| 193 | (7.3) | 356 | (13.4) | 0 | (0.0) | 409 | (15.4) | 2,651 |
| 254 | (11.1) | 0 | (0.0) | 0 | (0.0) | 259 | (11.3) | 2,294 |
| 133 | (6.2) | 79 | (3.7) | 0 | (0.0) | 634 | (29.6) | 2,138 |
| 143 | (6.8) | 271 | (12.9) | 0 | (0.0) | 426 | (20.3) | 2,096 |
| 189 | (9.0) | 138 | (6.6) | 0 | (0.0) | 54 | (2.6) | 2,092 |
| 390 | (18.8) | 0 | (0.0) | 0 | (0.0) | 330 | (15.9) | 2,075 |
| 85 | (4.2) | 0 | (0.0) | 35 | (1.7) | 57 | (2.8) | 2,008 |
| 718 | (43.9) | 240 | (14.7) | 100 | (6.1) | 414 | (25.3) | 1,637 |
| 0 | (0.0) | 28 | (1.7) | 0 | (0.0) | 96 | (6.0) | 1,608 |
| 346 | (22.2) | 156 | (10.0) | 0 | (0.0) | 95 | (6.1) | 1,559 |
| 280 | (18.2) | 63 | (4.1) | 0 | (0.0) | 68 | (4.4) | 1,536 |
| 86 | (7.2) | 189 | (16.0) | 0 | (0.0) | 122 | (10.3) | 1,182 |
| 117 | (10.0) | 42 | (3.6) | 0 | (0.0) | 139 | (11.9) | 1,163 |
| 45 | (4.1) | 0 | (0.0) | 0 | (0.0) | 34 | (3.1) | 1,097 |
| 0 | (0.0) | 0 | (0.0) | 0 | (0.0) | 0 | (0.0) | 1,086 |
| 0 | (0.0) | 0 | (0.0) | 0 | (0.0) | 0 | (0.0) | 658 |
| 3,145 | (0.2) | 1,881 | (0.1) | 134 | (0.0) | 3,306 | (0.2) | 29,966 |
| | | | | | | | | |
| 61 | (0.3) | 15 | (0.1) | 850 | (4.6) | 0 | (0.0) | 18,290 |
| 141 | (1.2) | 0 | (0.0) | 22 | (0.2) | 0 | (0.0) | 11,966 |
| 0 | (0.0) | 13 | (0.2) | 0 | (0.0) | 0 | (0.0) | 6,941 |
| 37 | (0.6) | 35 | (0.5) | 20 | (0.3) | 0 | (0.0) | 6,512 |
| 0 | (0.0) | 0 | (0.0) | 0 | (0.0) | 0 | (0.0) | 4,660 |
| 19 | (0.6) | 0 | (0.0) | 50 | (1.7) | 0 | (0.0) | 2,940 |
| 88 | (3.0) | 0 | (0.0) | 0 | (0.0) | 0 | (0.0) | 2,915 |
| 0 | (0.0) | 0 | (0.0) | 0 | (0.0) | 0 | (0.0) | 2,407 |
| 13 | (0.7) | 26 | (1.4) | 0 | (0.0) | 0 | (0.0) | 1,822 |
| 77 | (5.5) | 0 | (0.0) | 0 | (0.0) | 0 | (0.0) | 1,403 |
| 38 | (3.1) | 0 | (0.0) | 0 | (0.0) | 0 | (0.0) | 1,224 |
| 0 | (0.0) | 0 | (0.0) | 0 | (0.0) | 0 | (0.0) | 973 |
| 0 | (0.0) | 0 | (0.0) | 0 | (0.0) | 0 | (0.0) | 902 |
| 0 | (0.0) | 0 | (0.0) | 0 | (0.0) | 0 | (0.0) | 657 |
| 0 | (0.0) | 0 | (0.0) | 0 | (0.0) | 0 | (0.0) | 279 |
| 0 | (0.0) | 0 | (0.0) | 0 | (0.0) | 0 | (0.0) | 245 |
| 0 | (0.0) | 0 | (0.0) | 0 | (0.0) | 0 | (0.0) | 201 |
| 474 | (0.0) | 89 | (0.0) | 942 | (0.0) | 0 | (0.0) | 64,338 |
| | | | | | | | | |
| 0 | (0.0) | 0 | (0.0) | 0 | (0.0) | 208 | (0.8) | 27,430 |
| 0 | (0.0) | 0 | (0.0) | 0 | (0.0) | 0 | (0.0) | 22,447 |
| 1,537 | (8.8) | 0 | (0.0) | 434 | (2.5) | 604 | (3.4) | 17,535 |
| 0 | (0.0) | 0 | (0.0) | 93 | (1.1) | 93 | (1.1) | 8,599 |

| Institution | Natural Resources | Forest Resources | Crops | Animals |
|---|---|---|---|---|
| U. of Pennsylvania | 5 (0.1) | 0 (0.0) | 0 (0.0) | 7,478 (97.2) |
| Iowa State U. | 0 (0.0) | 0 (0.0) | 0 (0.0) | 6,134 (96.5) |
| U. of Georgia | 483 (8.3) | 0 (0.0) | 0 (0.0) | 5,156 (88.7) |
| Texas A&M U. | 905 (16.0) | 0 (0.0) | 0 (0.0) | 4,761 (84.0) |
| U. of Minnesota | 0 (0.0) | 0 (0.0) | 0 (0.0) | 5,245 (100.0) |
| U. of Missouri | 0 (0.0) | 0 (0.0) | 0 (0.0) | 4,748 (96.4) |
| U. of Florida | 65 (2.3) | 0 (0.0) | 0 (0.0) | 2,786 (97.7) |
| Auburn U. | 0 (0.0) | 0 (0.0) | 0 (0.0) | 2,666 (100.0) |
| Ohio State U. | 9 (0.4) | 0 (0.0) | 0 (0.0) | 2,157 (99.5) |
| Tufts U. | 0 (0.0) | 0 (0.0) | 0 (0.0) | 1,841 (99.3) |
| Louisiana State U. | 45 (3.2) | 0 (0.0) | 0 (0.0) | 1,353 (96.8) |
| North Carolina State U. | 43 (3.5) | 0 (0.0) | 0 (0.0) | 1,185 (96.5) |
| Oregon State U. | 98 (11.0) | 0 (0.0) | 0 (0.0) | 791 (89.0) |
| Washington State U. | 0 (0.0) | 0 (0.0) | 0 (0.0) | 195 (100.0) |
| Tuskegee U. | 0 (0.0) | 0 (0.0) | 0 (0.0) | 4 (100.0) |
| Subtotal | 6,178 (0.3) | 0 (0.0) | 0 (0.0) | 115,019 (5.8) |
| Grand total | 237,159 (11.9) | 165,811 (8.3) | 677,894 (34.0) | 555,732 (27.9) |

NOTE: Numbers in parentheses are percent of total research expenditures.
SOURCE: Data are from USDA Current Research Information System (CRIS).

| Institution | People, Communities, and Institutions, including Rural Development | Competition, Trade Adjustment, Price, and Income Policy | General Resource or Technology | Food Science and Human Nutrition | Total Research Expenditures |
|---|---|---|---|---|---|
| U. of Pennsylvania | 4 (0.0) | 0 (0.0) | 205 (2.7) | 0 (0.0) | 7,692 |
| Iowa State U. | 218 (3.4) | 0 (0.0) | 0 (0.0) | 3 (0.0) | 6,355 |
| U. of Georgia | 45 (0.8) | 0 (0.0) | 128 (2.2) | 0 (0.0) | 5,812 |
| Texas A&M U. | 0 (0.0) | 0 (0.0) | 0 (0.0) | 0 (0.0) | 5,667 |
| U. of Minnesota | 0 (0.0) | 0 (0.0) | 0 (0.0) | 0 (0.0) | 5,245 |
| U. of Missouri | 0 (0.0) | 174 (3.5) | 5 (0.1) | 0 (0.0) | 4,927 |
| U. of Florida | 0 (0.0) | 0 (0.0) | 0 (0.0) | 0 (0.0) | 2,851 |
| Auburn U. | 0 (0.0) | 0 (0.0) | 0 (0.0) | 0 (0.0) | 2,666 |
| Ohio State U. | 1 (0.0) | 0 (0.0) | 0 (0.0) | 0 (0.0) | 2,167 |
| Tufts U. | 0 (0.0) | 0 (0.0) | 14 (0.7) | 0 (0.0) | 1,855 |
| Louisiana State U. | 0 (0.0) | 0 (0.0) | 0 (0.0) | 0 (0.0) | 1,398 |
| North Carolina State U. | 0 (0.0) | 0 (0.0) | 0 (0.0) | 0 (0.0) | 1,228 |
| Oregon State U. | 0 (0.0) | 0 (0.0) | 0 (0.0) | 0 (0.0) | 888 |
| Washington State U. | 0 (0.0) | 0 (0.0) | 0 (0.0) | 0 (0.0) | 195 |
| Tuskegee U. | 0 (0.0) | 0 (0.0) | 0 (0.0) | 0 (0.0) | 4 |
| Subtotal | 1,805 (0.1) | 174 (0.0) | 879 (0.0) | 907 (0.0) | 124,961 |
| Grand total | 78,455 (3.9) | 86,528 (4.3) | 91,052 (4.6) | 99,102 (5.0) | 1,991,732 |

**APPENDIX TABLE 5** Forest Resources Research Expenditures (thousands of dollars) Ranked by Allocations to 1862 and 1890 Institutions and Forestry Schools, 1992

| Institution | Inventory and Appraisal of Forest Resources | | Timber Management | | Forest Protection | | Harvesting, Processing, and Marketing Forest Products | | Forest Watersheds, Soils, and Pollution | |
|---|---|---|---|---|---|---|---|---|---|---|
| *1862 State Agricultural Experiment Stations* | | | | | | | | | | |
| U. of California | 373 | (4.3) | 1,755 | (20.1) | 2,058 | (23.6) | 2,506 | (28.7) | 339 | (3.9) |
| Mississippi State U. | 476 | (6.7) | 708 | (10.0) | 159 | (2.2) | 5,207 | (73.3) | 63 | (0.9) |
| Virginia Polytechnic Institute and State U. | 366 | (5.4) | 942 | (13.9) | 398 | (5.9) | 2,971 | (43.9) | 532 | (7.9) |
| U. of Minnesota | 629 | (9.8) | 1,592 | (24.9) | 417 | (6.5) | 1,265 | (19.8) | 637 | (10.0) |
| U. of Florida | 84 | (1.3) | 2,842 | (45.2) | 640 | (10.2) | 17 | (0.3) | 471 | (7.5) |
| U. of Georgia | 0 | (0.0) | 1,473 | (27.2) | 381 | (7.1) | 1,649 | (30.5) | 292 | (5.4) |
| Texas A&M U. | 557 | (12.3) | 1,288 | (28.4) | 634 | (14.0) | 1,084 | (23.9) | 621 | (13.7) |
| Michigan State U. | 379 | (8.4) | 1,175 | (26.0) | 317 | (7.0) | 1,369 | (30.3) | 401 | (8.9) |
| Auburn U. | 65 | (1.7) | 1,480 | (38.7) | 572 | (14.9) | 861 | (22.5) | 565 | (14.8) |
| Pennsylvania State U. | 297 | (8.0) | 763 | (20.6) | 1,335 | (36.0) | 312 | (8.4) | 624 | (16.8) |
| U. of Maine | 195 | (5.4) | 1,342 | (37.0) | 88 | (2.4) | 576 | (15.9) | 686 | (18.9) |
| U. of Arizona | 275 | (8.6) | 157 | (4.9) | 220 | (6.9) | 0 | (0.0) | 707 | (22.1) |
| U. of Wisconsin | 269 | (9.8) | 693 | (25.1) | 568 | (20.6) | 640 | (23.2) | 12 | (0.5) |
| Purdue U. | 349 | (14.2) | 622 | (25.3) | 37 | (1.5) | 738 | (30.0) | 188 | (7.7) |
| West Virginia U. | 122 | (5.3) | 423 | (18.3) | 382 | (16.5) | 818 | (35.4) | 71 | (3.1) |
| Louisiana State U. | 0 | (0.0) | 622 | (27.6) | 250 | (11.1) | 394 | (17.5) | 263 | (11.7) |
| U. of Arkansas | 59 | (2.7) | 836 | (37.8) | 476 | (21.5) | 133 | (6.0) | 176 | (7.9) |
| Ohio State U. | 94 | (4.2) | 759 | (34.4) | 849 | (38.5) | 17 | (0.8) | 281 | (12.7) |
| U. of Tennessee | 0 | (0.0) | 887 | (43.1) | 391 | (19.0) | 265 | (12.9) | 0 | (0.0) |
| U. of Missouri | 0 | (0.0) | 329 | (18.9) | 141 | (8.1) | 62 | (3.6) | 455 | (26.2) |
| Iowa State U. | 0 | (0.0) | 609 | (35.4) | 313 | (18.2) | 323 | (18.7) | 421 | (24.5) |
| Colorado State U. | 0 | (0.0) | 417 | (24.5) | 103 | (6.0) | 194 | (11.4) | 0 | (0.0) |
| U. of Kentucky | 0 | (0.0) | 585 | (35.0) | 202 | (12.1) | 353 | (21.1) | 191 | (11.4) |
| U. of Illinois | 27 | (1.6) | 561 | (34.0) | 138 | (8.4) | 356 | (21.6) | 348 | (21.1) |
| Cornell U. | 413 | (25.9) | 361 | (22.6) | 465 | (29.1) | 145 | (9.1) | 50 | (3.1) |
| Oregon State U. | 291 | (18.3) | 566 | (35.6) | 240 | (15.1) | 0 | (0.0) | 266 | (16.8) |
| Washington State U. | 8 | (0.5) | 337 | (21.4) | 755 | (47.9) | 207 | (13.1) | 224 | (14.2) |
| U. of Maryland | 0 | (0.0) | 399 | (27.1) | 490 | (33.3) | 87 | (5.9) | 60 | (4.1) |
| Oklahoma State U. | 75 | (5.3) | 587 | (42.0) | 0 | (0.0) | 188 | (13.5) | 216 | (15.5) |
| U. of Massachusetts | 0 | (0.0) | 386 | (34.2) | 586 | (51.8) | 0 | (0.0) | 61 | (5.4) |
| Utah State U. | 29 | (2.7) | 277 | (25.4) | 192 | (17.6) | 28 | (2.6) | 120 | (11.0) |
| U. of Alaska | 7 | (0.6) | 754 | (69.9) | 0 | (0.0) | 37 | (3.4) | 11 | (1.0) |
| U. of Nebraska | 0 | (0.0) | 292 | (28.9) | 52 | (5.1) | 83 | (8.2) | 179 | (17.7) |
| U. of Hawaii | 0 | (0.0) | 547 | (54.9) | 159 | (15.9) | 197 | (19.8) | 27 | (2.8) |
| U. of Connecticut | 4 | (0.4) | 218 | (22.5) | 508 | (52.4) | 0 | (0.0) | 215 | (22.2) |
| New Mexico State U. | 146 | (15.8) | 494 | (53.5) | 64 | (6.9) | 0 | (0.0) | 39 | (4.2) |
| North Carolina State U. | 59 | (6.4) | 386 | (41.9) | 387 | (42.0) | 50 | (5.4) | 17 | (1.8) |
| Montana State U. | 0 | (0.0) | 0 | (0.0) | 31 | (3.5) | 0 | (0.0) | 0 | (0.0) |
| North Dakota State U. | 0 | (0.0) | 0 | (0.0) | 174 | (19.9) | 0 | (0.0) | 0 | (0.0) |
| U. of New Hampshire | 63 | (8.5) | 142 | (19.2) | 51 | (6.9) | 4 | (0.5) | 189 | (25.5) |
| Rutgers—The State U. | 161 | (22.3) | 241 | (33.3) | 58 | (8.0) | 110 | (15.2) | 35 | (4.9) |
| Kansas State U. | 153 | (24.2) | 354 | (56.0) | 78 | (12.3) | 0 | (0.0) | 0 | (0.0) |
| U. of Idaho | 0 | (0.0) | 1 | (0.1) | 0 | (0.0) | 215 | (36.0) | 0 | (0.0) |
| U. of Nevada | 127 | (25.6) | 259 | (52.2) | 38 | (7.6) | 0 | (0.0) | 67 | (13.4) |
| U. of Wyoming | 43 | (8.8) | 0 | (0.0) | 66 | (13.5) | 45 | (9.2) | 33 | (6.8) |
| U. of Vermont | 0 | (0.0) | 94 | (19.4) | 389 | (80.6) | 0 | (0.0) | 0 | (0.0) |
| South Dakota State U. | 72 | (22.2) | 136 | (41.8) | 0 | (0.0) | 0 | (0.0) | 28 | (8.7) |
| U. of Rhode Island | 200 | (71.7) | 79 | (28.3) | 0 | (0.0) | 0 | (0.0) | 0 | (0.0) |
| U. of Delaware | 0 | (0.0) | 0 | (0.0) | 161 | (57.9) | 0 | (0.0) | 0 | (0.0) |
| U. of Puerto Rico | 0 | (0.0) | 0 | (0.0) | 0 | (0.0) | 0 | (0.0) | 150 | (100.0) |
| U. of The Virgin Islands | 0 | (0.0) | 94 | (100.0) | 0 | (0.0) | 0 | (0.0) | 0 | (0.0) |
| U. of Guam | 0 | (0.0) | 0 | (0.0) | 30 | (40.4) | 0 | (0.0) | 0 | (0.0) |
| U. of the District of Columbia | 0 | (0.0) | 0 | (0.0) | 0 | (0.0) | 0 | (0.0) | 58 | (100.0) |
| Clemson U. | 0 | (0.0) | 0 | (0.0) | 0 | (0.0) | 0 | (0.0) | 42 | (85.4) |
| Subtotal | 6,467 | (3.9) | 29,864 | (18.0) | 16,041 | (9.7) | 23,504 | (14.2) | 10,434 | (6.3) |

| Forest, Range, Wildlife, and Fisheries Habitat Development | Forest Recreation, Aesthetics and other Landscape Values | Alternative Uses of Land | Technical Assistance | Trees, Forests, and Forest Products | Forest and Range Resources | Total Research Expenditures |
|---|---|---|---|---|---|---|
| 1,057 (12.1) | 95 (1.1) | 258 (3.0) | 260 (3.0) | 0 (0.0) | 34 (0.4) | 8,735 |
| 211 (3.0) | 0 (0.0) | 256 (3.6) | 0 (0.0) | 26 (0.4) | 0 (0.0) | 7,105 |
| 1,042 (15.4) | 0 (0.0) | 375 (5.5) | 31 (0.5) | 108 (1.6) | 0 (0.0) | 6,765 |
| 112 (1.8) | 487 (7.6) | 1,012 (15.9) | 228 (3.6) | 6 (0.1) | 0 (0.0) | 6,385 |
| 2,003 (31.9) | 7 (0.1) | 0 (0.0) | 223 (3.5) | 0 (0.0) | 0 (0.0) | 6,288 |
| 1,278 (23.6) | 8 (0.2) | 78 (1.4) | 78 (1.4) | 173 (3.2) | 0 (0.0) | 5,410 |
| 0 (0.0) | 98 (2.2) | 154 (3.4) | 27 (0.6) | 78 (1.7) | 0 (0.0) | 4,541 |
| 0 (0.0) | 189 (4.2) | 75 (1.7) | 576 (12.7) | 41 (0.9) | 0 (0.0) | 4,520 |
| 62 (1.6) | 12 (0.3) | 82 (2.1) | 15 (0.4) | 75 (2.0) | 41 (1.1) | 3,829 |
| 128 (3.4) | 86 (2.3) | 162 (4.4) | 0 (0.0) | 0 (0.0) | 0 (0.0) | 3,706 |
| 222 (6.1) | 90 (2.5) | 430 (11.9) | 0 (0.0) | 0 (0.0) | 0 (0.0) | 3,630 |
| 601 (18.8) | 234 (7.3) | 410 (12.8) | 6 (0.2) | 0 (0.0) | 588 (18.4) | 3,198 |
| 4 (0.1) | 16 (0.6) | 149 (5.4) | 349 (12.6) | 0 (0.0) | 57 (2.1) | 2,758 |
| 102 (4.1) | 257 (10.4) | 111 (4.5) | 18 (0.7) | 22 (0.9) | 15 (0.6) | 2,461 |
| 189 (8.2) | 223 (9.7) | 85 (3.7) | 0 (0.0) | 0 (0.0) | 0 (0.0) | 2,315 |
| 375 (16.7) | 25 (1.1) | 146 (6.5) | 175 (7.8) | 0 (0.0) | 0 (0.0) | 2,250 |
| 91 (4.1) | 49 (2.2) | 62 (2.8) | 330 (14.9) | 0 (0.0) | 0 (0.0) | 2,211 |
| 1 (0.1) | 10 (0.5) | 36 (1.6) | 158 (7.2) | 0 (0.0) | 0 (0.0) | 2,205 |
| 374 (18.2) | 0 (0.0) | 134 (6.5) | 6 (0.3) | 0 (0.0) | 0 (0.0) | 2,057 |
| 344 (19.8) | 0 (0.0) | 254 (14.6) | 115 (6.6) | 38 (2.2) | 0 (0.0) | 1,736 |
| 0 (0.0) | 32 (1.9) | 23 (1.3) | 0 (0.0) | 0 (0.0) | 0 (0.0) | 1,721 |
| 301 (17.7) | 47 (2.8) | 560 (32.9) | 80 (4.7) | 0 (0.0) | 0 (0.0) | 1,701 |
| 176 (10.5) | 0 (0.0) | 0 (0.0) | 165 (9.9) | 0 (0.0) | 0 (0.0) | 1,672 |
| 6 (0.4) | 25 (1.5) | 158 (9.6) | 8 (0.5) | 0 (0.0) | 24 (1.4) | 1,651 |
| 4 (0.2) | 1 (0.1) | 27 (1.7) | 130 (8.2) | 0 (0.0) | 0 (0.0) | 1,595 |
| 206 (13.0) | 0 (0.0) | 19 (1.2) | 0 (0.0) | 0 (0.0) | 0 (0.0) | 1,589 |
| 12 (0.8) | 2 (0.1) | 24 (1.5) | 7 (0.4) | 0 (0.0) | 1 (0.0) | 1,578 |
| 0 (0.0) | 132 (8.9) | 16 (1.1) | 287 (19.5) | 0 (0.0) | 0 (0.0) | 1,472 |
| 0 (0.0) | 0 (0.0) | 0 (0.0) | 331 (23.7) | 0 (0.0) | 0 (0.0) | 1,397 |
| 62 (5.5) | 0 (0.0) | 35 (3.1) | 0 (0.0) | 0 (0.0) | 0 (0.0) | 1,130 |
| 317 (29.0) | 30 (2.7) | 42 (3.8) | 56 (5.1) | 0 (0.0) | 0 (0.0) | 1,093 |
| 231 (21.4) | 23 (2.2) | 16 (1.4) | 0 (0.0) | 0 (0.0) | 0 (0.0) | 1,078 |
| 16 (1.5) | 0 (0.0) | 0 (0.0) | 357 (35.3) | 33 (3.3) | 0 (0.0) | 1,011 |
| 37 (3.7) | 0 (0.0) | 0 (0.0) | 30 (3.0) | 0 (0.0) | 0 (0.0) | 997 |
| 0 (0.0) | 2 (0.2) | 22 (2.3) | 0 (0.0) | 0 (0.0) | 0 (0.0) | 968 |
| 99 (10.7) | 29 (3.2) | 2 (0.2) | 50 (5.4) | 0 (0.0) | 0 (0.0) | 923 |
| 0 (0.0) | 0 (0.0) | 0 (0.0) | 23 (2.5) | 0 (0.0) | 0 (0.0) | 921 |
| 214 (24.0) | 11 (1.2) | 19 (2.1) | 601 (67.2) | 0 (0.0) | 18 (2.0) | 894 |
| 0 (0.0) | 164 (18.7) | 0 (0.0) | 538 (61.4) | 0 (0.0) | 0 (0.0) | 875 |
| 262 (35.5) | 0 (0.0) | 29 (3.9) | 0 (0.0) | 0 (0.0) | 0 (0.0) | 740 |
| 0 (0.0) | 10 (1.3) | 0 (0.0) | 11 (1.6) | 98 (13.5) | 0 (0.0) | 724 |
| 35 (5.5) | 12 (2.0) | 0 (0.0) | 0 (0.0) | 0 (0.0) | 0 (0.0) | 632 |
| 0 (0.0) | 0 (0.0) | 0 (0.0) | 236 (39.6) | 145 (24.3) | 0 (0.0) | 597 |
| 0 (0.0) | 0 (0.0) | 0 (0.0) | 0 (0.0) | 0 (0.0) | 6 (1.2) | 497 |
| 301 (61.7) | 0 (0.0) | 0 (0.0) | 0 (0.0) | 0 (0.0) | 0 (0.0) | 488 |
| 0 (0.0) | 0 (0.0) | 0 (0.0) | 0 (0.0) | 0 (0.0) | 0 (0.0) | 482 |
| 0 (0.0) | 47 (14.3) | 0 (0.0) | 43 (13.0) | 0 (0.0) | 0 (0.0) | 326 |
| 0 (0.0) | 0 (0.0) | 0 (0.0) | 0 (0.0) | 0 (0.0) | 0 (0.0) | 280 |
| 117 (42.1) | 0 (0.0) | 0 (0.0) | 0 (0.0) | 0 (0.0) | 0 (0.0) | 277 |
| 0 (0.0) | 0 (0.0) | 0 (0.0) | 0 (0.0) | 0 (0.0) | 0 (0.0) | 150 |
| 0 (0.0) | 0 (0.0) | 0 (0.0) | 0 (0.0) | 0 (0.0) | 0 (0.0) | 94 |
| 0 (0.0) | 0 (0.0) | 0 (0.0) | 44 (59.6) | 0 (0.0) | 0 (0.0) | 74 |
| 0 (0.0) | 0 (0.0) | 0 (0.0) | 0 (0.0) | 0 (0.0) | 0 (0.0) | 58 |
| 0 (0.0) | 0 (0.0) | 0 (0.0) | 7 (14.6) | 0 (0.0) | 0 (0.0) | 49 |
| 10,591 (6.4) | 2,453 (1.5) | 5,259 (3.2) | 5,597 (3.4) | 841 (0.5) | 784 (0.5) | 111,836 |

| Institution | Inventory and Appraisal of Forest Resources | Timber Management | Forest Protection | Harvesting, Processing, and Marketing Forest Products | Forest Watersheds, Soils, and Pollution |
|---|---|---|---|---|---|
| *1890 Colleges and Universities* | | | | | |
| Tennessee State U. | 0 (0.0) | 0 (0.0) | 204 (100.0) | 0 (0.0) | 0 (0.0) |
| Kentucky State College | 0 (0.0) | 117 (68.7) | 0 (0.0) | 0 (0.0) | 0 (0.0) |
| Florida A&M U. | 0 (0.0) | 0 (0.0) | 0 (0.0) | 0 (0.0) | 0 (0.0) |
| Alabama A&M U. | 0 (0.0) | 30 (100.0) | 0 (0.0) | 0 (0.0) | 0 (0.0) |
| Subtotal | 0 (0.0) | 147 (0.1) | 204 (0.1) | 0 (0.0) | 0 (0.0) |
| *Forestry Schools* | | | | | |
| State U. of New York | 107 (0.9) | 4,381 (36.4) | 306 (2.5) | 5,156 (42.9) | 830 (6.9) |
| Oregon State U. | 135 (1.3) | 4,171 (40.1) | 453 (4.4) | 3,616 (34.8) | 904 (8.7) |
| North Carolina State U. | 269 (4.3) | 2,972 (48.0) | 55 (0.9) | 2,022 (32.6) | 490 (7.9) |
| U. of Washington | 365 (5.9) | 2,461 (40.1) | 242 (3.9) | 747 (12.2) | 1,321 (21.5) |
| Michigan Technological U. | 1,332 (29.3) | 1,470 (32.4) | 30 (0.7) | 1,010 (22.2) | 517 (11.4) |
| U. of Montana | 773 (29.1) | 716 (26.9) | 57 (2.1) | 91 (3.4) | 245 (9.2) |
| Stephen F. Austin State U. | 505 (21.8) | 430 (18.5) | 260 (11.2) | 147 (6.3) | 248 (10.7) |
| U. of Idaho | 135 (6.1) | 1,166 (52.9) | 197 (8.9) | 143 (6.5) | 362 (16.4) |
| Clemson U. | 105 (6.4) | 786 (47.8) | 120 (7.3) | 268 (16.3) | 122 (7.4) |
| Northern Arizona U. | 292 (22.0) | 175 (13.2) | 76 (5.7) | 13 (1.0) | 527 (39.7) |
| U. of Michigan | 139 (11.9) | 556 (47.4) | 210 (17.9) | 78 (6.6) | 73 (6.2) |
| U. of Vermont | 0 (0.0) | 94 (10.3) | 94 (10.4) | 0 (0.0) | 450 (49.4) |
| Southern Illinois U. | 22 (3.5) | 303 (49.2) | 0 (0.0) | 167 (27.1) | 61 (9.9) |
| Colorado State U. | 0 (0.1) | 0 (0.0) | 0 (0.0) | 536 (99.9) | 0 (0.0) |
| Louisiana Tech U. | 0 (0.0) | 158 (56.8) | 41 (14.8) | 45 (16.0) | 35 (12.5) |
| Humboldt State U. | 71 (29.0) | 151 (61.7) | 7 (2.9) | 0 (0.0) | 0 (0.0) |
| California Polytechnic State U. | 151 (75.0) | 50 (25.0) | 0 (0.0) | 0 (0.0) | 0 (0.0) |
| Subtotal | 4,402 (2.7) | 20,040 (12.1) | 2,150 (1.3) | 14,037 (8.5) | 6,184 (3.7) |
| Total | 10,869 (6.6) | 50,051 (30.2) | 18,395 (11.1) | 37,541 (22.6) | 16,617 (10.0) |

NOTE: Number in parentheses is percent of total expenditures. Overall, 67.4%, 0.3%, and 32.2% of total forest resources research expenditures are allocated to 1862s, 1890s, and forestry schools, respectively.

SOURCE: Data are from USDA Current Research Information System (CRIS).

| Institution | Forest, Range, Wildlife, and Fisheries Habitat Development | Forest Recreation, Aesthetics and other Landscape Values | Alternative Uses of Land | Technical Assistance | Trees, Forests, and Forest Products | Forest and Range Resources | Total Research Expenditures |
|---|---|---|---|---|---|---|---|
| *1890 Colleges and Universities* | | | | | | | |
| Tennessee State U. | 0 (0.0) | 0 (0.0) | 0 (0.0) | 0 (0.0) | 0 (0.0) | 0 (0.0) | 204 |
| Kentucky State College | 0 (0.0) | 0 (0.0) | 0 (0.0) | 53 (31.3) | 0 (0.0) | 0 (0.0) | 170 |
| Florida A&M U. | 0 (0.0) | 0 (0.0) | 0 (0.0) | 154 (100.0) | 0 (0.0) | 0 (0.0) | 154 |
| Alabama A&M U. | 0 (0.0) | 0 (0.0) | 0 (0.0) | 0 (0.0) | 0 (0.0) | 0 (0.0) | 30 |
| Subtotal | 0 (0.0) | 0 (0.0) | 0 (0.0) | 207 (0.1) | 0 (0.0) | 0 (0.0) | 557 |
| *Forestry Schools* | | | | | | | |
| State U. of New York | 208 (1.7) | 48 (0.4) | 375 (3.1) | 0 (0.0) | 199 (1.7) | 411 (3.4) | 12,021 |
| Oregon State U. | 502 (4.8) | 563 (5.4) | 42 (0.4) | 16 (0.2) | 0 (0.0) | 0 (0.0) | 10,402 |
| North Carolina State U. | 130 (2.1) | 49 (0.8) | 209 (3.4) | 0 (0.0) | 0 (0.0) | 0 (0.0) | 6,197 |
| U. of Washington | 329 (5.4) | 173 (2.8) | 365 (5.9) | 69 (1.1) | 0 (0.0) | 70 (1.1) | 6,143 |
| Michigan Technological U. | 182 (4.0) | 0 (0.0) | 0 (0.0) | 0 (0.0) | 0 (0.0) | 0 (0.0) | 4,540 |
| U. of Montana | 488 (18.4) | 264 (9.9) | 5 (0.2) | 0 (0.0) | 0 (0.0) | 18 (0.7) | 2,657 |
| Stephen F. Austin State U. | 0 (0.0) | 317 (13.7) | 412 (17.8) | 0 (0.0) | 0 (0.0) | 0 (0.0) | 2,319 |
| U. of Idaho | 119 (5.4) | 0 (0.0) | 83 (3.7) | 0 (0.0) | 0 (0.0) | 0 (0.0) | 2,204 |
| Clemson U. | 101 (6.1) | 31 (1.9) | 111 (6.7) | 0 (0.0) | 0 (0.0) | 0 (0.0) | 1,644 |
| Northern Arizona U. | 33 (2.5) | 61 (4.6) | 151 (11.4) | 0 (0.0) | 0 (0.0) | 0 (0.0) | 1,327 |
| U. of Michigan | 1 (0.1) | 13 (1.1) | 104 (8.9) | 0 (0.0) | 0 (0.0) | 0 (0.0) | 1,175 |
| U. of Vermont | 74 (8.1) | 39 (4.3) | 143 (15.7) | 0 (0.0) | 0 (0.0) | 17 (1.9) | 911 |
| Southern Illinois U. | 0 (0.0) | 63 (10.3) | 0 (0.0) | 0 (0.0) | 0 (0.0) | 0 (0.0) | 616 |
| Colorado State U. | 0 (0.0) | 0 (0.0) | 0 (0.0) | 0 (0.0) | 0 (0.0) | 0 (0.0) | 536 |
| Louisiana Tech U. | 0 (0.0) | 0 (0.0) | 0 (0.0) | 0 (0.0) | 0 (0.0) | 0 (0.0) | 279 |
| Humboldt State U. | 16 (6.4) | 0 (0.0) | 0 (0.0) | 0 (0.0) | 0 (0.0) | 0 (0.0) | 245 |
| California Polytechnic State U. | 0 (0.0) | 0 (0.0) | 0 (0.0) | 0 (0.0) | 0 (0.0) | 0 (0.0) | 201 |
| Subtotal | 2,182 (1.3) | 1,622 (1.0) | 2,000 (1.2) | 85 (0.1) | 199 (0.1) | 516 (0.3) | 53,418 |
| Total | 12,773 (7.7) | 4,075 (2.5) | 7,259 (4.4) | 5,889 (3.6) | 1,041 (0.6) | 1,300 (0.8) | 165,811 |

**APPENDIX TABLE 6** Allocation (percent) of Commodity Research Expenditures among Specific Crops and Animals Research, by 1862 Institution, 1992

| Institution | Citrus and Tropical /Sub-tropical Fruit | Deciduous and Small Fruits and Edible Tree Nuts | Potatoes and Vegetables | Ornamentals and Turf | Corn and Grain Sorghum | Rice | Wheat and Other Small Grains | Pasture and Forage Crops |
|---|---|---|---|---|---|---|---|---|
| Texas A&M U. | 1.1 | 2.5 | 7.2 | 4.6 | 10.8 | 3.2 | 6.1 | 4.5 |
| U. of Florida | 20.4 | 3.8 | 21.6 | 10.2 | 2.6 | 1.4 | 1.0 | 6.6 |
| U. of California | 11.6 | 21.4 | 21.2 | 4.2 | 1.0 | 1.1 | 5.6 | 3.0 |
| North Carolina State U. | 0.0 | 9.4 | 12.7 | 5.0 | 9.2 | 2.3 | 2.5 | 2.7 |
| Iowa State U. | 0.0 | 0.7 | 3.2 | 2.0 | 22.3 | 0.0 | 3.0 | 2.5 |
| U. of Georgia | 0.0 | 9.7 | 11.2 | 6.7 | 5.3 | 0.1 | 4.6 | 8.3 |
| U. of Nebraska | 0.0 | 0.4 | 4.2 | 3.2 | 11.9 | 0.0 | 6.0 | 2.9 |
| U. of Minnesota | 0.0 | 3.0 | 12.6 | 3.9 | 12.7 | 0.8 | 8.2 | 5.7 |
| Kansas State U. | 0.0 | 1.4 | 2.7 | 1.9 | 12.2 | 0.6 | 29.1 | 4.7 |
| Louisiana State U. | 0.7 | 3.2 | 10.0 | 2.3 | 2.9 | 9.3 | 3.2 | 7.1 |
| Purdue U. | 0.0 | 3.3 | 9.3 | 2.5 | 23.7 | 1.1 | 4.8 | 2.2 |
| U. of Wisconsin | 0.0 | 3.4 | 16.1 | 0.3 | 6.2 | 0.3 | 2.3 | 5.7 |
| U. of Illinois | 0.1 | 2.4 | 4.5 | 3.6 | 19.6 | 0.0 | 2.8 | 3.7 |
| Washington State U. | 0.0 | 24.3 | 18.6 | 2.9 | 1.1 | 1.0 | 19.6 | 4.1 |
| U. of Arkansas | 0.0 | 6.1 | 6.3 | 1.2 | 3.2 | 15.8 | 4.6 | 2.1 |
| Michigan State U. | 0.0 | 16.7 | 19.2 | 5.7 | 2.0 | 0.5 | 3.9 | 2.4 |
| Ohio State U. | 0.0 | 13.8 | 9.5 | 8.7 | 7.5 | 0.0 | 2.7 | 2.9 |
| Cornell U. | 0.1 | 4.1 | 28.0 | 5.3 | 6.1 | 2.7 | 5.1 | 6.8 |
| Oregon State U. | 0.0 | 22.5 | 19.4 | 1.8 | 0.3 | 0.0 | 13.7 | 3.1 |
| Virginia Polytechnic Institute and State U. | 0.0 | 4.9 | 7.3 | 6.9 | 5.9 | 0.0 | 2.6 | 2.7 |
| Pennsylvania State U. | 0.0 | 11.6 | 13.9 | 8.1 | 6.6 | 0.0 | 2.6 | 3.0 |
| U. of Arizona | 0.4 | 6.9 | 18.4 | 2.9 | 8.9 | 0.0 | 1.8 | 8.1 |
| Oklahoma State U. | 0.0 | 6.1 | 9.3 | 2.5 | 0.7 | 0.0 | 15.6 | 5.1 |
| U. of Tennessee | 0.0 | 4.3 | 9.4 | 7.8 | 3.7 | 0.0 | 2.3 | 9.1 |
| Auburn U. | 0.0 | 6.6 | 8.1 | 6.8 | 2.8 | 0.0 | 2.3 | 6.2 |
| North Dakota State U. | 0.0 | 0.2 | 14.2 | 0.4 | 3.5 | 0.0 | 32.4 | 2.6 |
| U. of Missouri | 0.1 | 2.8 | 1.1 | 3.5 | 10.1 | 1.6 | 4.5 | 9.5 |
| Clemson U. | 0.0 | 8.5 | 12.1 | 9.0 | 3.3 | 0.2 | 5.1 | 4.9 |
| U. of Kentucky | 0.0 | 2.2 | 9.7 | 3.8 | 4.8 | 0.0 | 2.2 | 10.9 |
| Mississippi State U. | 0.0 | 5.4 | 5.1 | 6.1 | 6.4 | 4.6 | 1.2 | 7.1 |
| U. of Hawaii | 20.2 | 5.9 | 24.2 | 13.8 | 3.3 | 0.3 | 5.8 | 2.0 |
| U. of Idaho | 0.0 | 3.5 | 32.0 | 0.7 | 0.2 | 0.0 | 25.2 | 4.2 |
| Colorado State U. | 0.0 | 5.5 | 11.3 | 3.5 | 9.8 | 0.0 | 21.0 | 7.0 |
| U. of Maryland | 0.0 | 6.1 | 10.5 | 4.8 | 4.9 | 0.1 | 7.3 | 7.4 |
| Montana State U. | 0.0 | 0.0 | 5.8 | 0.0 | 2.4 | 0.0 | 30.5 | 9.1 |
| U. of Puerto Rico | 17.7 | 1.4 | 27.9 | 1.9 | 0.1 | 1.3 | 0.0 | 12.5 |
| Geneva AES | 0.5 | 59.2 | 34.6 | 3.0 | 0.7 | 0.4 | 0.7 | 0.9 |
| South Dakota State U. | 0.0 | 0.9 | 1.2 | 1.3 | 8.7 | 0.0 | 16.0 | 3.9 |
| Rutgers—The State U., Cook College | 0.0 | 17.5 | 20.2 | 15.5 | 1.5 | 0.0 | 1.5 | 0.2 |
| Utah State U. | 0.0 | 5.4 | 0.7 | 1.3 | 0.9 | 0.0 | 15.7 | 11.1 |
| New Mexico State U. | 0.0 | 1.5 | 21.4 | 3.0 | 4.0 | 0.0 | 3.1 | 19.2 |
| U. of Delaware | 0.1 | 0.1 | 10.9 | 2.9 | 15.8 | 0.2 | 0.8 | 0.5 |
| U. of Massachusetts | 0.0 | 25.4 | 16.1 | 4.7 | 5.3 | 0.0 | 0.0 | 14.0 |
| U. of Maine | 0.0 | 24.2 | 48.2 | 0.6 | 1.7 | 0.0 | 1.3 | 2.1 |
| U. of Wyoming | 0.0 | 0.4 | 0.3 | 0.0 | 0.0 | 0.0 | 1.8 | 9.7 |
| U. of Connecticut | 0.0 | 0.7 | 3.0 | 12.5 | 0.0 | 0.0 | 0.1 | 2.6 |
| West Virginia U. | 0.0 | 14.7 | 4.8 | 1.3 | 0.0 | 0.0 | 0.2 | 12.4 |
| New Haven AES | 0.0 | 10.6 | 32.8 | 10.9 | 1.2 | 0.0 | 1.7 | 1.4 |
| U. of New Hampshire | 0.0 | 14.7 | 13.1 | 0.9 | 4.8 | 0.0 | 0.0 | 0.0 |

| Cotton and Cotton Seed | Soybeans and Other Oilseed and Oil Crops | Peanuts | Tobacco | Sugar Crops | Miscel-laneous and New Crops | Poultry | Cattle | | Swine | Sheep and Wool, Other Animals, Bees and Honey |
|---|---|---|---|---|---|---|---|---|---|---|
| | | | | | | | Beef | Dairy | | |
| 10.5 | 2.0 | 3.2 | 0.0 | 2.6 | 1.3 | 4.8 | 20.2 | 5.5 | 1.0 | 8.8 |
| 0.1 | 2.3 | 2.0 | 0.2 | 1.5 | 0.6 | 2.0 | 10.0 | 6.6 | 3.5 | 3.5 |
| 2.9 | 1.7 | 0.0 | 0.0 | 0.8 | 0.5 | 3.8 | 4.3 | 5.4 | 0.5 | 10.9 |
| 0.8 | 4.6 | 4.8 | 14.0 | 0.0 | 0.2 | 9.6 | 5.7 | 8.1 | 5.1 | 3.2 |
| 0.0 | 16.9 | 0.1 | 0.0 | 0.0 | 0.6 | 3.3 | 14.0 | 7.1 | 21.8 | 2.5 |
| 2.8 | 8.3 | 11.2 | 2.1 | 0.0 | 0.3 | 8.2 | 12.5 | 5.0 | 3.6 | 0.3 |
| 0.0 | 7.0 | 0.0 | 0.0 | 1.0 | 1.6 | 2.8 | 36.6 | 5.0 | 10.5 | 6.9 |
| 0.0 | 7.3 | 0.0 | 0.1 | 1.3 | 1.3 | 7.6 | 6.6 | 13.1 | 13.2 | 2.4 |
| 0.0 | 6.4 | 0.0 | 0.0 | 0.0 | 0.0 | 3.1 | 21.9 | 4.8 | 6.1 | 5.2 |
| 7.9 | 10.0 | 0.1 | 0.0 | 7.7 | 0.6 | 3.6 | 12.9 | 12.2 | 2.2 | 4.1 |
| 0.0 | 7.1 | 0.0 | 0.5 | 1.3 | 3.1 | 5.7 | 10.0 | 8.2 | 12.8 | 4.4 |
| 0.0 | 2.8 | 0.0 | 0.1 | 0.0 | 0.4 | 4.1 | 8.3 | 37.5 | 5.6 | 6.8 |
| 0.1 | 15.3 | 0.0 | 0.2 | 0.9 | 0.6 | 4.3 | 10.8 | 12.4 | 14.8 | 3.9 |
| 0.0 | 1.2 | 0.0 | 0.4 | 0.0 | 5.8 | 2.6 | 6.5 | 7.5 | 2.1 | 2.1 |
| 10.3 | 15.7 | 0.2 | 0.0 | 0.3 | 0.2 | 16.5 | 10.6 | 4.1 | 1.4 | 1.4 |
| 0.0 | 3.3 | 0.0 | 0.3 | 1.4 | 9.7 | 4.8 | 7.8 | 12.7 | 6.3 | 3.2 |
| 0.1 | 8.5 | 0.0 | 1.0 | 0.3 | 0.0 | 8.8 | 6.5 | 14.4 | 8.5 | 6.7 |
| 0.0 | 0.6 | 0.0 | 0.0 | 0.1 | 3.6 | 7.9 | 3.6 | 19.5 | 1.2 | 5.3 |
| 0.0 | 3.4 | 0.0 | 0.6 | 0.4 | 4.9 | 4.0 | 7.2 | 4.2 | 0.8 | 13.8 |
| | | | | | | | | | | |
| 0.2 | 5.8 | 3.9 | 3.5 | 0.8 | 0.4 | 8.5 | 14.6 | 13.0 | 8.8 | 10.2 |
| 0.0 | 1.6 | 0.0 | 0.2 | 0.0 | 1.4 | 12.7 | 3.6 | 22.3 | 7.7 | 4.6 |
| 15.6 | 0.6 | 0.0 | 1.3 | 0.7 | 3.1 | 0.7 | 10.6 | 11.2 | 1.5 | 7.4 |
| 5.4 | 1.5 | 4.3 | 0.0 | 0.0 | 0.3 | 1.7 | 27.6 | 5.5 | 4.5 | 9.9 |
| 4.3 | 7.5 | 0.0 | 5.0 | 0.0 | 0.4 | 2.2 | 13.4 | 21.9 | 7.1 | 1.7 |
| 7.5 | 5.5 | 5.9 | 0.0 | 0.0 | 0.0 | 18.6 | 13.7 | 5.1 | 9.3 | 1.5 |
| 0.0 | 17.7 | 0.0 | 0.0 | 1.5 | 1.6 | 0.0 | 12.0 | 4.3 | 4.9 | 4.4 |
| 1.6 | 14.0 | 0.0 | 0.1 | 0.0 | 0.8 | 4.1 | 13.4 | 9.6 | 13.4 | 9.8 |
| 6.2 | 9.1 | 0.1 | 5.7 | 0.0 | 0.2 | 10.2 | 10.3 | 8.4 | 3.6 | 3.2 |
| 0.0 | 7.5 | 0.0 | 10.5 | 0.0 | 0.1 | 2.6 | 13.8 | 7.8 | 5.1 | 18.9 |
| 13.2 | 11.6 | 0.0 | 0.0 | 0.0 | 3.4 | 9.7 | 11.3 | 10.9 | 3.1 | 0.8 |
| 0.0 | 0.1 | 0.0 | 0.8 | 1.2 | 6.6 | 2.9 | 7.6 | 3.8 | 0.2 | 1.3 |
| 0.0 | 8.5 | 0.0 | 0.0 | 4.3 | 0.0 | 0.2 | 11.2 | 3.5 | 0.2 | 6.4 |
| 0.0 | 1.3 | 0.0 | 0.0 | 0.1 | 0.0 | 0.1 | 31.2 | 1.6 | 0.1 | 7.3 |
| 0.0 | 8.5 | 0.0 | 4.0 | 0.0 | 0.3 | 21.8 | 4.7 | 14.0 | 2.2 | 3.5 |
| 0.0 | 4.8 | 0.0 | 0.0 | 1.7 | 1.1 | 0.0 | 26.2 | 1.8 | 5.9 | 10.6 |
| 0.0 | 0.1 | 0.0 | 0.0 | 16.5 | 12.3 | 0.0 | 0.9 | 7.4 | 0.0 | 0.0 |
| 0.0 | 0.0 | 0.0 | 0.0 | 0.0 | 0.0 | 0.0 | 0.0 | 0.0 | 0.0 | 0.0 |
| 0.0 | 7.3 | 0.0 | 0.0 | 0.0 | 0.9 | 1.2 | 29.5 | 11.0 | 12.8 | 5.5 |
| | | | | | | | | | | |
| 0.4 | 3.1 | 0.2 | 2.3 | 1.3 | 3.2 | 12.1 | 0.0 | 2.0 | 2.3 | 16.8 |
| 0.0 | 0.0 | 0.0 | 0.0 | 0.0 | 0.1 | 6.7 | 9.8 | 25.6 | 0.7 | 21.8 |
| 11.4 | 4.5 | 2.2 | 0.0 | 0.4 | 2.3 | 1.1 | 19.2 | 0.1 | 0.0 | 6.5 |
| 0.2 | 7.7 | 0.0 | 0.0 | 0.0 | 1.6 | 43.3 | 0.8 | 10.7 | 4.0 | 0.3 |
| 0.0 | 0.0 | 0.0 | 0.8 | 0.0 | 0.0 | 5.4 | 2.1 | 23.0 | 0.0 | 3.1 |
| 0.0 | 0.0 | 0.0 | 0.0 | 0.0 | 0.0 | 1.3 | 0.0 | 19.6 | 0.0 | 1.1 |
| 0.0 | 0.7 | 0.0 | 0.0 | 3.3 | 0.0 | 0.0 | 36.4 | 2.2 | 0.2 | 44.6 |
| 0.0 | 0.0 | 0.0 | 0.0 | 0.0 | 0.9 | 19.1 | 3.6 | 30.9 | 0.0 | 25.9 |
| 0.0 | 0.0 | 0.0 | 0.0 | 0.0 | 1.1 | 18.1 | 14.1 | 21.3 | 0.7 | 11.1 |
| 0.0 | 0.0 | 0.0 | 29.9 | 0.0 | 1.6 | 0.8 | 0.5 | 2.4 | 0.5 | 5.8 |
| 0.0 | 5.3 | 0.0 | 0.0 | 0.0 | 2.4 | 12.4 | 1.4 | 27.4 | 10.0 | 7.6 |

| Institution | Citrus and Tropical/ Subtropical Fruit | Deciduous and Small Fruits and Edible Tree Nuts | Potatoes and Vegetables | Ornamentals and Turf | Corn and Grain Sorghum | Rice | Wheat and Other Small Grains | Pasture and Forage Crops |
|---|---|---|---|---|---|---|---|---|
| U. of Guam | 9.9 | 1.2 | 63.6 | 11.6 | 0.0 | 0.0 | 0.0 | 0.0 |
| U. of Vermont | 0.0 | 18.5 | 3.2 | 3.4 | 1.5 | 0.0 | 0.0 | 9.7 |
| U. of Alaska | 0.0 | 2.8 | 15.0 | 9.7 | 0.0 | 0.0 | 15.8 | 13.4 |
| U. of Nevada | 0.0 | 0.0 | 1.2 | 9.5 | 1.9 | 0.0 | 3.8 | 9.9 |
| U. of Rhode Island | 0.0 | 7.2 | 27.5 | 43.4 | 0.0 | 0.0 | 0.0 | 0.0 |
| U. of the Virgin Islands | 12.1 | 0.0 | 40.4 | 5.5 | 0.0 | 0.0 | 0.0 | 15.9 |
| American Samoa Community College | 0.0 | 0.0 | 100.0 | 0.0 | 0.0 | 0.0 | 0.0 | 0.0 |
| Northern Marianas College | 0.0 | 0.0 | 100.0 | 0.0 | 0.0 | 0.0 | 0.0 | 0.0 |
| U. of the District of Columbia | 0.0 | 0.0 | 56.4 | 34.9 | 8.7 | 0.0 | 0.0 | 0.0 |
| Total | 1.6 | 7.6 | 18.5 | 5.8 | 5.1 | 0.8 | 6.1 | 5.5 |

NOTE: Institutions are ranked by decreasing order of reported expenditures for research on selected commodities or commodity groupings. The total commodity research expenditures at all 1862 institutions is $978.06 million. Research that is not commodity specific, usually basic research applicable to multiple crops, is not included in this table.

SOURCE: Data are from USDA Current Research Information Systems (CRIS)

| Cotton and Cotton Seed | Soybeans and Other Oilseed and Oil Crops | Peanuts | Tobacco | Sugar Crops | Miscel-laneous and New Crops | Poultry | Cattle | | Swine | Sheep and Wool, Other Animals, Bees and Honey |
|---|---|---|---|---|---|---|---|---|---|---|
| | | | | | | | Beef | Dairy | | |
| 0.0 | 0.0 | 0.0 | 0.0 | 0.0 | 1.2 | 12.3 | 0.0 | 0.0 | 0.0 | 0.0 |
| 0.0 | 0.0 | 0.0 | 0.0 | 0.0 | 0.0 | 4.1 | 0.0 | 52.4 | 0.0 | 7.2 |
| 0.0 | 0.0 | 0.0 | 0.0 | 0.0 | 0.0 | 0.0 | 8.7 | 21.0 | 7.9 | 5.6 |
| 0.2 | 0.0 | 0.0 | 0.8 | 0.0 | 1.9 | 0.0 | 40.1 | 18.9 | 1.1 | 10.9 |
| 0.0 | 13.0 | 0.0 | 0.0 | 0.0 | 0.0 | 3.5 | 0.0 | 3.5 | 0.0 | 1.9 |
| 0.0 | 0.0 | 0.0 | 0.0 | 0.0 | 0.0 | 0.0 | 0.0 | 0.0 | 0.0 | 26.1 |
| 0.0 | 0.0 | 0.0 | 0.0 | 0.0 | 0.0 | 0.0 | 0.0 | 0.0 | 0.0 | 0.0 |
| 0.0 | 0.0 | 0.0 | 0.0 | 0.0 | 0.0 | 0.0 | 0.0 | 0.0 | 0.0 | 0.0 |
| 0.0 | 0.0 | 0.0 | 0.0 | 0.0 | 0.0 | 0.0 | 0.0 | 0.0 | 0.0 | 0.0 |
| 1.8 | 4.7 | 0.7 | 1.5 | 0.9 | 1.5 | 6.0 | 10.5 | 10.6 | 4.1 | 6.9 |

# References


Alston, J. M., P. G. Pardey, and H. O. Carter, eds. 1994. Valuing UC Agricultural Research and Extension. Agricultural Issues Center, Publication No. VR-1. Davis: University of California, Division of Agricultural and Natural Resources.

Alston, J. M., and P. G. Pardey. 1995. Making Science Pay: Economics of Financing, Organizing and Managing Public-Sector Agricultural R&D. Washington, D.C.: American Enterprise Institute.

Alston, J. M., G. W. Norton, and P. G. Pardey. 1995. Science Under Scarcity: Principles and Practice for Agricultural Research Evaluation and Priority Setting. Ithaca, N.Y.: Cornell University Press.

Caswell, M., K. O. Fuglie, and C. A. Klotz. 1994. Agricultural Biotechnology: An Economic Perspective, Agricultural Economics Report No. 687, Economic Research Service. Washington, D.C.: U.S. Department of Agriculture.

Christy, R. D., and L. Williamson, eds. 1992. A Century of Service: Land-Grant Colleges and Universities, 1890-1990. New Brunswick, N.J.: Transaction Publishers.

Council on Food, Agricultural and Resource Economics. 1994. Industrialization of U.S. Agriculture: Policy, Research, and Education Needs. April 1994. Photocopy.

Executive Office of the President. 1994. Economic Report of the President: Transmitted to Congress February 1994. Washington, D.C.: U.S. Government Printing Office.

Executive Office of the President. 1994. Budget of the United States Government. Washington, D.C.: U.S. Government Printing Office.

Executive Office of the President. 1995. Budget of the United States Government. Washington, D.C.: U.S. Government Printing Office.

Experiment Station Committee on Organization and Policy. 1994. Opportunities to Meet Changing Needs: Research on Food, Agriculture, and Natural Resources. Washington, D.C.: U.S. Department of Agriculture.

Food and Agricultural Education Information System. 1994. Fall 1993 Enrollment in Agriculture, Renewable Natural Resources and Forestry: A Combined Report. College Station: Texas A&M University.

Frazão, E. 1995. The American Diet: Health and Economic Consequences. Agriculture Information Bulletin No. 711. Washington, D.C.: U.S. Department of Agriculture, Economic Research Service.

Hoppe, R. A. 1994. Farming Operations and Households in Farming Areas: A Closer Look, Agricultural Economics Report No. 685. Washington, D.C.: U.S. Department of Agriculture, Economic Research Service.

Huffman, W. E., and R. E. Evenson. 1989. Supply and demand functions for multiproduct U.S. cash grain farms: Biases caused by research and other policies. Am. J. Agric. Econ. 71:761-773.

Huffman, W. E., and R. E. Evenson. 1993. Science for Agriculture: A Long-Term Perspective. Ames: Iowa State University Press.

Kerr, N. A. 1987. The Legacy: A Centennial History of the State Agricultural Experiment Stations, 1887-1987. Columbia: Missouri Agricultural Experiment Station .

Leones, J., G. Schulter, and G. Goldman. 1994. Redefining agriculture in interindustry analysis. Am. J. Agric. Econ. 76(5):1123-1129.

McGrath Morris, P., L. Neuhauser, and C. Campbell. 1992. Food security in rural America: A study of the availability and costs of food. J. Nutr. Ed. 24(1):52S-58S.

Myers, J. H. 1991. Rethinking the Outlook of Colleges Whose Roots Have Been in Agriculture. Davis: University of California.

National Association of State Universities and Land-Grant Colleges. 1995. The Unique Land-Grant System Working for People, Food, Agriculture, Environment Through Teaching, Research and Extension. Washington, D.C.: National Association of State Universities and Land-Grant Colleges.

National Research Council. 1989. Investing in Research: A Proposal to Strengthen the Agricultural, Food, and Environmental System. Washington, D.C.: National Academy Press.

National Research Council. 1994. Investing in the National Research Initiative: An Update of the Competitive Grants Program in the U.S. Department of Agriculture. Washington, D.C.: National Academy Press.

National Science Foundation. 1991. Federal Support to Universities, Colleges, and Nonprofit Institutions: Fiscal Year 1991. Arlington, Va.: National Science Foundation.

National Science Foundation. 1993. Interagency Program Announcement: DOE/NSF/USDA Joint Program on Collaborative Research in Plant Biology . Arlington, Va.: National Science Foundation.

Peterson, R., and N. Brooks. 1993. The Changing Concentration of U.S. Agricultural Production During the 20th Century: 14th Annual Report to the Congress on the Status of the Family Farm, Agriculture Information Bulletin No. 671. Washington, D.C.: U.S. Department of Agriculture, Economic Research Service.

Powell, D. S., et al. 1993. Forest Resources of the United States, 1992. USDA Forest Service, General Technical Report RM-234. Washington, D.C.: U.S. Forest Service.

Rasmussen, W. D. 1989. Taking the University to the People: Seventy-Five Years of Cooperative Extension. Ames: Iowa State University Press.

Sommer, J. E., and F. K. Hines. 1991. Diversity in U.S. Agriculture: A New Delineation by Farming Characteristics, Agricultural Economics Report No. 646. Washington, D.C.: U.S. Department of Agriculture, Economic Research Service.

Troutt, D. D. 1993. The Thin Red Line: How the Poor Still Pay More. San Francisco: West Coast Regional Office of Consumers Union.

U.S. Department of Agriculture. 1990a. An Analysis of the Timber Situation in the United States, 1989-2040: A Technical Document Supporting the 1989 USDA Forest Service RPA Assessment. General Technical Report RM-199. Fort Collins, Colo.: Rocky Mountain Forest and Range Experiment Station, Forest Service, U.S. Department of Agriculture.

U.S. Department of Agriculture. 1990b. State Financial Summary. Washington, D.C.: U.S. Department of Agriculture.

U.S. Department of Agriculture. 1992. Agricultural Statistics, 1992. Washington, D.C.: U.S. Department of Agriculture.

U.S. Department of Agriculture. 1993a. Cooperative State Research Service, Manual of Classification of Agricultural and Forestry Research. Washington, D.C.: U.S. Department of Agriculture.

U.S. Department of Agriculture. 1993b. 1993-94 Directory of Professional Workers in State Agricultural Experiment Stations and Other Cooperating State Institutions. Washington, D.C.: U.S. Department of Agriculture.

U.S. Department of Agriculture, Economic Research Service, Natural Resources and Environment Division. 1994. Agricultural Resources and Economic Indicators. Agricultural Handbook No. 705. Washington, D.C: U.S. Department of Agriculture.

U.S. Department of Education. 1991. Digest of Education Statistics. Washington, D.C.: U.S. Department of Education.

U.S. Department of Education. 1993. Digest of Education Statistics 1993. Washington, D.C.: U.S. Department of Education.

# ABOUT THE AUTHORS

**Anthony S. Earl** (*chair*) has been a partner in the law firm of Quarles & Brady, Madison, Wisconsin, since 1987. Prior to that he had been governor of Wisconsin (1983-1986). Earl received his J.D. degree from the University of Chicago in 1961. An advocate of environmental and civic responsibility, as governor, he successfully advanced through the legislature a significant number of initiatives in the areas of education, equal opportunity, economic development, and protection of the environment. Earl also served as secretary of the departments of natural resources and of administration for the state of Wisconsin and is a co-founder of the Center for Clean Air Policy and a board member of the Great Lakes Protection Fund, Common Cause, Resources for the Future, and other environmental and civic organizations.

**R. Lee Baldwin**, affiliated with the University of California since 1963, is Sesnon Professor of Animal Science at the University of California, Davis. He has been the recipient of both Guggenheim and Fulbright fellowships. Baldwin earned his M.S. degree in dairy nutrition and Ph.D. degree in biochemistry and nutrition from Michigan State University. His primary areas of research interest include modeling ruminant digestion and animal metabolism, nutritional energetics, and the physiology of lactation. Baldwin was elected to the National Academy of Sciences in 1993.

**John C. Gordon** is Pinchot Professor and the acting director of the Yale Institute of Biospheric Studies at Yale University. He received his B.S. degree and his Ph.D. degree in plant physiology from Iowa State University. Gordon's primary areas of expertise are plant physiology and silviculture, and his research includes work in tree physiology and ecology, especially biological nitrogen fixation. A member of the Society of American Foresters as well as other organizations, he is actively working to develop research methodologies and science and policy interactions.

**Gordon E. Guyer**, director of the Michigan Department of Agriculture, originally retired from Michigan State University (MSU) in 1986 after a distinguished 33-year career in agriculture education and natural resources development. In 1998 he was recruited to return as professor emeritus and vice president for government affairs. In 1992 he was recruited to return as president of MSU, a position he held until 1993. Guyer was formerly the director of the Michigan Department of Natural Resources. His B.S., M.S., and Ph.D. degrees were earned in entomology at MSU. He has served as consultant to governments worldwide on agriculture and natural resource issues.

**Fred Harrison, Jr.**, since 1982, has been administrator and director of the Cooperative Extension Program at the School of Agriculture, Home Economics, and Allied Programs, Fort Valley State College, Fort Valley, Georgia. He also serves on the executive committee of USDA's Joint Council on Food and Agricultural Sciences and has chaired the Extension Committee on Organization and Policy. Harrison earned his M.Ed. degree from the University of Georgia and his Ph.D. degree in agricultural education and administration from Ohio State University. His area of research is agricultural extension education and administration.

**Edward A. Hiler** is vice chancellor and dean of Agriculture and Life Sciences at Texas A&M University and director of the Texas Agricultural Experiment Station. Prior to that he was head of Texas A&M University's Department of Agricultural Engineering (1974–1988). Hiler earned his B.S., M.S., and Ph.D. degrees in agricultural engineering from Ohio State University. He serves as a consultant to the Office of Technology Assessment, the Office of Water Research and Technology, and several U.S. and western European universities regarding environmental quality and the future direction of agricultural engineering. He is a past-president of the American Society of Agricultural Engineers and in 1987 was elected to the National Academy of Engineering.

**Marlyn L. Jorgensen** is a partner in Jorg-Anna Farms and president and C.E.O. of Timberlane Hogs, Ltd., Garrison, Iowa. He is president of the Iowa Producers Co-Op and a board member of Sunrise Energy. He is also past-president and chair of the American Soybean Association, past-director of Iowa Farm Bureau Federation, and past-coordinator of Benton Rural Development Group. Jorgensen received his degree in animal science from Iowa State University. He is most actively involved in farming economics in relation to national economic indicators and government policy.

**Daryl B. Lund** became dean of the College of Agriculture and Life Sciences at Cornell University in August 1995. Prior to that, he was the executive dean and executive director of Cook College and of the New Jersey Experiment Station at Rutgers—The State University of New Jersey. After receiving his Ph.D. degree in food science and chemical engineering from the University of Wisconsin in 1968, Lund joined the Department of Food Science as a professor of food engineering. He served as chair of that department (1984–1987) until he moved to Rutgers as chair of the Food Science Department. His research expertise includes food process engineering with special emphasis on simultaneous heat and mass transfer, energy and food processing, and development of microwave-assisted heat and mass transfer operations.

**Thomas F. Malone** is a Distinguished University Scholar at North Carolina State University and director of the Sigma Xi Center, Research Triangle Park, North Carolina. Malone has held tenured faculty appointments at the Massachusetts Institute of Technology and the University of Connecticut where he was dean of the graduate school. He has served as president of the Meteorological Society, the American Geophysical Union, and of the Scientific Honor Society Sigma Xi. Elected to the National Academy of Sciences in 1968, Malone served as Foreign Secretary of the National Academy of Sciences from 1978 until 1982. His Sc.D. was earned from the Massachusetts Institute of Technology in 1946. Malone's primary area of research is sustainable human development.

**Mortimer H. Neufville** is acting vice president for academic affairs and, since 1983, has been dean of School of Agricultural Sciences and the 1890 research director at the University of Maryland, Eastern Shore. He is also the associate director of the Maryland Agricultural Experiment Station. Neufville's responsibilities include supervising thirteen academic departments and a comprehensive program encompassing many aspects of agriculture in domestic and international research and education programs in food and agricultural sciences. Neufville earned his M.S. and Ph.D. degrees in animal science from the University of Florida, Gainesville.

**Elizabeth D. Owens** is manager of product registration at ISK Biosciences Corporation, Mentor, Ohio. Her responsibilities include supervising product registrations for crop protection products and managing new and existing applications and reregistration compliance. Prior to joining ISK, Owens was manager of commercial development and regulatory affairs for BioTechnica International, Inc., Cambridge, Massachusetts. Ownes has held research positions in industry and at the University of Massachusetts, both Boston and Amherst campuses. Owens earned her B.S. degree from the University of Idaho, M.S. degree from Iowa State University, and Ph.D. degree from the University of Massachusetts.

**C. Alan Pettibone** is superintendent of the Western Washington Research and Extension Centers (at Puyallup, Vancouver, Long Beach, and Mt. Vernon), Washington State University (WSU). When named to the committee, he was assistant director of Cooperative Extension for Agriculture and Natural Resources at WSU. Pettibone received his Ph.D. degree in Agricultural Engineering from Cornell University and has extensive experience in agriculture and natural resource issues from both a technical and policy viewpoint, having served in a number of administrative positions at WSU and as director of the Washington State Department of Agriculture for almost a decade.

**Allen Rosenfeld** is senior vice president for programs and acting co-director at Public Voice for Food and Health Policy, a Washington, D.C. based consumer group. He has been responsible for Public Voice's legislative and regulatory work on food safety, food labeling, biotechnology, pesticide policy, and sustainable agriculture. Before joining Public Voice, Rosenfeld conducted field research in Guatemala on expansion of "nontraditional" exports and was an assistant professor in the Agricultural Management Department of the California Polytechnic State University (1982–1986) and, while there, also was coordinator of the International Agricultural Development Program. He received his Ph.D. degree in agricultural economics from Cornell University.

**Charles F. Saul**, prior to his retirement in 1995, was president, C.E.O., and general manager of Agway, Inc., Syracuse, New York, a food marketing cooperative owned by 91,000 farmer-stockholders. Saul joined Agway in 1954. Following military service, he returned to Agway and began his advance from district manager through vice president, to group vice president, executive vice president and chief operations officer, assistant general manager, and finally general manager and president. Having built a career in agribusiness management, Saul now serves on the boards of directors for several agribusiness corporations and advocacy organizations in New York state. Saul is a graduate of Cornell University.

**G. Edward Schuh** is dean of the Hubert H. Humphrey Institute of Public Affairs at the University of Minnesota, St. Paul. Prior to that he was director of Agriculture and Rural Development for the World Bank in Washington, D.C., (1984–1987) and head of the Department of Agricultural and Applied Economics, University of Minnesota (1979–1984). Schuh holds an M.S. degree from Michigan State University and M.A. and Ph.D.

degrees from the University of Chicago. His career has included serving in various academic capacities at Purdue University (1959–1979), as program advisor to the Ford Foundation in Brazil (1966–1972), as senior staff economist on President Ford's Council of Economic Advisors (1974–1975), and as USDA deputy undersecretary for International Affairs and Commodity Programs (1978–1979).

**George E. Seidel, Jr.,** is professor of reproductive physiology at Colorado State University. He received his B.S. degree from The Pennsylvania State University and M.S. and Ph.D. degrees in reproductive physiology from Cornell University. He has served as a research fellow at Harvard Medical School; professor in the Department of Physiology, at Colorado State; was a visiting associate professor in the Biology Department at Yale University; and was visiting scientist at the Whitehead Institute. Seidel's research includes superovulation and embryo transfer, in vitro oocyte maturation and fertilization, cryopreservation of livestock embryos, and embryo microsurgery. Seidel was elected to the National Academy of Sciences in 1992.

**Jo Ann Doke Smith** is a founding partner of the consulting firm of Smith Associates, Texas and Florida, and has extensive experience in both public and private sectors of the agriculture industry. She has served as USDA's assistant secretary for Marketing and Inspection Services, president of the National Cattlemen's Association, and a member of the governor's Task Force on the Future of Florida Agriculture. Ms. Smith is an active member of the boards of directors of the Iowa Beef Producers, Inc., and Purina Mills, Inc., and is involved in agricultural marketing as it relates to government policy.

**Katherine R. Smith** is director of the Policy Studies Program at the Henry A. Wallace Institute for Alternative Agriculture, Greenbelt, Maryland. The interdisciplinary program she leads assesses the implications of policy alternatives for the sustainability of agricultural systems. Her 14 year tenure with USDA's Economic Research Service included positions as associate director of Resources and Technology Division, chief of the Western Hemisphere Branch, and acting administrator. Smith earned her Ph.D. in agricultural and resource economics from the University of Maryland. Smith's principal areas of expertise are agricultural and resource policies and the relationship between agricultural production and environmental quality.

**James B. Wyngaarden** recently retired from a distinguished career that included holding the following positions: chair of the Department of Medicine at Duke University (1967–1982) and associate vice chancellor of Duke University, director of the National Institutes of Health (1982–1989), Foreign Secretary of the National Academy of Sciences and the Institute of Medicine (1990–1994), and, concurrently, director of the Human Genome Organization (1990–1994). Dr. Wyngaarden received his M.D. degree from the University of Michigan and was elected to the Institute of Medicine in 1973 and the National Academy of Sciences in 1974; he is a fellow of the Royal College of Physicians of London; and a member of the American Society of Clinical Investigators.

**Elisabeth A. Zinser** became chancellor of the University of Kentucky in July 1995 after having served as president of the University of Idaho. She was also the chair of the Committee on Outreach and Technology Transfer for the National Association of State Universities and Land-Grant Colleges. Zinser has served as administrator of the School of Medicine at the University of Washington, Seattle; professor and dean of the College of Nursing at the University of North Dakota; vice chancellor for academic affairs at the University of North Carolina, Greensboro; and president of Gallaudet University in Washington, D.C. She earned M.S. degrees from the University of California at San Francisco and from the Massachusetts Institute of Technology and received her Ph.D. degree from the University of California at Berkeley.

**James J. Zuiches** became dean of the College of Agriculture and Home Economics at Washington State University in July 1995, following his work as program director for Food Systems and Rural Development with the W.K. Kellogg Foundation, Battle Creek, Michigan. When named to the committee, he was director of the Washington State University Agricultural Research Center and associate dean of the College of Agriculture and Home Economics. Zuiches has also been a research administrator with Cornell University and with the National Science Foundation. He received his M.S. and Ph.D. degrees in sociology from the University of Wisconsin, Madison. Zuiches' primary area of research is in population studies, particularly migration models, labor force dynamics, and rural development.

## ABOUT THE STAFF

**Nicole Ballenger** is a senior agricultural economist at the U.S. Department of Agriculture's Economic Research Service (ERS). She directs the land grant study under the terms of an Intergovernmental Personnel Act Agreement between the ERS and the National Research Council. Ballenger has a Ph.D. degree in agricultural economics from the University of California at Davis. Her areas of economics research include U.S. agricultural and food policy, international trade, agriculture in developing countries, and linkages between trade and the environment. Prior to directing the land grant study she was a branch chief in the former Resources and Technology Division of ERS. During 1990-1991 she was senior staff economist for agriculture and trade for the President's Council of Economic Advisers.

**Carla Carlson** is director of communications at the Board on Agriculture, where she is responsible for the writing, editing, review, production, and dissemination of reports. She coordinated public forums in the states as part of the land grant study. Prior to coming to the National Research Council, she was science correspondent for the U.S. Information Agency's wire service and associate editor of *SciQuest* magazine. She is an officer of the D.C. Science Writers' Association and a member of its board of directors. She has degrees in journalism and biology from South Dakota State University.

**Viola Horek** is administrative assistant at the Board on Agriculture and senior project assistant for its Committee on the Future of the Land Grant Colleges of Agriculture. From 1990 to 1993 she worked for the city of Stuttgart, Germany as urban planner, and earlier was employed by the Department of Defense in Germany. She received her M.A. degree in architecture and urban planning from the University of Stuttgart.

**Diby M. Kouadio** is a research associate for the Board on Agriculture's study of the land grant colleges of agriculture. He earned his M.S. degree in agricultural economics from the University of Illinois at Urbana-Champaign and his Ph.D. degree in agricultural economics from The Pennsylvania State University. Kouadio's economics research interests relate to policy intervention in world commodity markets. He is a member of the American Agricultural Economics Association and the National Society for Minorities in Agriculture, Natural Resources and Related Sciences.

**Janet Overton** has edited the Board on Agriculture's major reports since 1991. Earlier, she was production editor with scientific publisher Marcel Dekker, Inc., and production editor at the *Proceedings of the National Academy of Sciences*. She has also consulted and freelanced for a number of publishing houses including Random House, D. Van Nostrand, Springer-Verlag, and Prentice-Hall. Overton holds a M.F.A. degree in play writing from Columbia University. She is a published playwright and reader for professional theaters.

# Index

Italicized page numbers indicate that the subject appears in a table.

## RECENT PUBLICATIONS OF THE BOARD ON AGRICULTURE

### Policy and Resources

Investing in the National Research Initiative: An Update of the Competitive Grants Program in the U.S. Department of Agriculture (1994), 64 pp., ISBN 0-309-05235-1
Rangeland Health: New Methods to Classify, Inventory, and Monitor Rangelands (1994), 180 pp., ISBN 0-309-04879-6
Soil and Water Quality: An Agenda for Agriculture (1993), 516 pp., ISBN 0-309-04933-4
Pesticides in the Diets of Infants and Children (1993), 408 pp., ISBN 0-309-04875-3
Managing Global Genetic Resources: Agricultural Crop Issues and Policies (1993), 450 pp., ISBN 0-309-04430-8
Managing Global Genetic Resources: Livestock (1993), 294 pp., ISBN 0-309-04394-8
Sustainable Agriculture and the Environment in the Humid Tropics (1993), 720 pp., ISBN 0-309-04749-8
Agriculture and the Undergraduate: Proceedings (1992), 296 pp., ISBN 0-309-04682-3
Water Transfers in the West: Efficiency, Equity, and the Environment (1992), 320 pp., ISBN 0-309-04528-2
Managing Global Genetic Resources: Forest Trees (1991), 244 pp., ISBN 0-309-04034-5
Managing Global Genetic Resources: The U.S. National Plant Germplasm System (1991), 198 pp., ISBN 0-309-04390-5
Sustainable Agriculture Research and Education in the Filed: A Proceedings (1991), 448 pp., ISBN 0-309-04578-9
Toward Sustainability: A Plan for Collaborative Research on Agriculture and Natural Resource Management (1991), 164, pp., ISBN 0-309-04540-1
Investing in Research: A Proposal to Strengthen the Agricultural, Food, and Environmental System (1989), 156 pp., ISBN 0-309-04127-9
Alternative Agriculture (1989), 464 pp., ISBN 0-309-03985-1
Understanding Agriculture: New Directions for Education (1988), 80 pp., ISBN 0-309-03936-3
Designing Foods: Animal Product Options in the Marketplace (1988), 394 pp., ISBN 0-309-03798-0; ISBN 0-309-03795-6 (pbk)
Agricultural Biotechnology: Strategies for National Competitiveness (1987), 224 pp., ISBN 0-309-03745-X
Regulating Pesticides in Food: The Delaney Paradox (1987), 288 pp., ISBN 0-309-03746-8
Pesticide Resistance: Strategies and Tactics for Management (1986), 480 pp., ISBN 0-309-03627-5
Pesticides and Groundwater Quality: Issues and Problems in Four States (1986), 136 pp., ISBN 0-309-03676-3
Soil Conservation: Assessing the National Resources Inventory, Volume 1 (1986), 134 pp., ISBN 0-309-03649-9; Volume 2 (1986), 314 pp., ISBN 0-309-03675-5
New Directions for Biosciences Research in Agriculture: High-Reward Opportunities (1985), 122 pp., ISBN 0-309-03542-2
Genetic Engineering of Plants: Agricultural Research Opportunities and Policy Concerns (1984), 96 pp., ISBN 0-309-03434-5

### Nutrient Requirements of Domestic Animals Series and Related Titles

Nutrient Requirements of Laboratory Animals, Fourth Revised Edition (1995), 174 pp., ISBN 0-309-05126-6
Nutrient Requirements of Poultry, Ninth Revised Edition (1994), 156 pp., ISBN 0-309-04892-3
Nutrient Requirements of Fish (1993), 108 pp., ISBN 0-309-04891-5
Nutrient Requirements of Horses, Fifth Revised Edition (1989), 128 pp., ISBN 0-309-03989-4; diskette included
Nutrient Requirements of Dairy Cattle, Sixth Revised Edition, Update 1989 (1989), 168 pp., ISBN 0-309-03826-X; diskette included Nutrient Requirements of Swine, Ninth Revised Edition (1988), 96 pp., ISBN 0-309-03779-4
Vitamin Tolerance of Animals (1987), 105 pp., ISBN 0-309-03728-X Predicting Feed Intake of Food-Producing Animals (1986), 95 pp., ISBN 0-309-03695-X
Nutrient Requirements of Cats, Revised Edition (1986), 87 pp., ISBN 0-309-03682-8
Nutrient Requirements of Dogs, Revised Edition (1985), 79 pp., ISBN 0-309-03496-5
Nutrient Requirements of Sheep, Sixth Revised Edition (1985), 106 pp., ISBN 0-309-03596-1
Nutrient Requirements of Beef Cattle, Sixth Revised Edition (1984), 90 pp., ISBN 0-309-03447-7

Further information, additional titles (prior to 1984), and prices are available from the National Academy Press, 2101 Constitution Avenue, NW, Washington, DC 20418, 202/334-3313 (information only); 800/624-6242 (orders only); 202/334-2451 (fax).

